APR 1 8 1979

D1455589

388.44
REED R
NEW YORK ELEVATED

Port Washington Public Library
Port Washington, New York
Phone: 883-4400

4770100481

470 F

APR 1 8 1979

The New York Elevated

ALSO BY ROBERT C. REED:

Train Wrecks
The Streamline Era

The New York Elevated

Robert C. Reed

South Brunswick and New York: A. S. Barnes and Company
London: Thomas Yoseloff Ltd

© 1978 by A. S. Barnes and Co., Inc.

A. S. Barnes and Co., Inc.
Cranbury, New Jersey 08512

Thomas Yoseloff Ltd
Magdalen House
136-148 Tooley Street
London SE1 2TT, England

Library of Congress Cataloging in Publication Data
Reed, Robert Carroll, 1937–
  The New York elevated.

  Includes index.
  1. Railroads, Elevated—New York (City) I. Title.
TF841.N48R43   388.4′4′097471   77-84581
ISBN 0-498-02138-6

PRINTED IN THE UNITED STATES OF AMERICA

For John and Marian Reed

Contents

Acknowledgments

The elevated railroads of New York City have been an all-absorbing subject to me since the spring of 1972, when Robert Vogel, Curator of Civil Engineering for the Smithsonian Institution, first suggested that I might be interested in writing a book about them. When I set out on the project, I knew rather little about the el, aside from those obvious things known to most people. But soon the bits and pieces began to fall together, and the el began to breathe and rattle again for me.

Although I have written this book for the general reader, I have felt obliged to wrestle with the fundamental technology of the elevated, and because I am not an engineer, the technical side of my research has often been demanding. To temper the metallic ring of the technology, I have sought to place the el where it belongs—amidst the personal and social history of the times. To do this I talked to as many people as I could find who remembered riding and seeing the elevated trains. Often such conversations would reveal valuable clues that I could later pursue on my own. Almost everyone who knew the el had cogent statements to make about it, and they were nearly universally positive comments, I might add. To round out the basic information about the el, I painstakingly searched book by book the vast local-history section of the Library of Congress, gleaning where I could any intimate, personal observations of the subject throughout its history.

My research trips to New York City provided not only useful information, but helped enormously to sustain my enthusiasm for the project through the tedious months of hunting, sorting, writing, and rewriting. Just walking along the routes of the old trestles was exciting to me. I will never forget the nearly metaphysical experience of wandering one foggy March night from a

Greenwich Village restaurant along deserted Greenwich Street to my room in Chelsea. At one point at Little West 12th Street, where the tracks once turned into Ninth Avenue, the el seemed for a moment to reappear—then it quickly dissolved into the darkness.

I am deeply grateful to the many people who assisted me in preparing this book: Wilson G. Duprey, Curator of the Map and Print Room at the New York Historical Society; George T. F. Rahilly of Fort Lauderdale, Florida; Edward B. Watson of New York City; the staff of the Science and Technology Division of the New York Public Library; and the Electronic Railroaders Assn. for their useful rosters and chronologies. Charlotte La Rue, Photo Librarian at the Museum of the City of New York, was generous of her time in digging out old views and clippings of the el from her collections.

I am very much indebted to E. Alfred Seibel of Claymont, Delaware, for his help in answering my many questions. He kindly shared his own extensive collection on the New York Elevated and allowed me to read his own manuscript on the subject.

My hearty thanks also to Jerry Kearns and Milton Kaplan at the Prints and Photos Division of the Library of Congress and to John White, Curator of Transportation for the Smithsonian Institution. Marilyn Wandrus of the National Park Service gave me many valuable tips on the intricacies of picture research, and Marty Hublitz was good enough to take time out from his own busy schedule to help with the photography. Anthony Adams was superb, without his help and encouragement I could never have completed the book.

Finally, I wish to thank all my friends in Washington and New York whose recollections of the elevated were helpful in recapturing the memory of it.

The New York Elevated

1 New York in the 1860s: The Need for the El

New York's elevated railroads had their beginnings in the mid 1860s, when travel conditions in the overcrowded city on the tip of Manhattan Island had grown so intolerable as to call for some better, more rapid transportation system than slow, animal-drawn buses and stages. With a population of nearly a million people and the largest city in the country at the close of the Civil War, New York was packed with people and with more immigrants arriving from Europe daily, many of whom never moved onward but made the city their home. New York was becoming so crowded in 1865 that it was dangerous to cross Broadway, the city's busiest street, on foot because of the fierce congestion of wagons, drays, omnibuses, horsecars, and carts. And the primitive, animal-powered, public-transit system was hopelessly inadequate to meet the needs of the population.

Still, with all of its obvious transportation problems, many New Yorkers in the mid 1860s had other thoughts on their minds than the slowness of city travel. So deeply engrossed had most people been in the throes of their Civil War that there was, generally speaking, a preoccupation with optimism and lighter thoughts in the years immediately after 1864 and Appomattox. It would be another year or so before they would turn their thoughts to the building of elevated railroads. It was a time when people wondered at Barnum's American Museum and Niblo's Garden, when women wore hoopskirts, and when Maggie Mitchell was making herself famous in *Fanchon* and Negro minstrelsy was in its glory, when the Academy of Music was the home of Grand Opera, and Brignoli, Zucchi, and Massimiliani sang in *Il Trovatore* and other early Verdi operas, when Theodore Thomas was building up his famous orchestra and giving "symphony soirees" at Irving Hall. It was the year that saw the re-election of Abraham Lincoln as president and the admission of Nevada as a state to the Union.

The city had become a rich metropolis, for in spite of the war its people had plenty of money and lavished it on all sorts of pleasures. The newspapers commented on the brilliance of the winter season, the extravagance of the women, and the expensive restaurants, of which Delmonico's set the fashion. In one editorial the *Times* reported that women of fashion thought nothing of paying a hundred dollars for a bonnet, yet criticized the omnibus companies for raising their fares from six to ten cents and the theater for preparing to increase the price of admission beyond fifty cents. Yet while New York was a rich city—commerical, financial, and artistic capital of the American continent—it was a poor city as well. And in the wrong wind the lower island stank with the exhalations of slaughterhouses, gasworks, manure heaps, decaying garbage, and the festering bodies of dead animals. The human death rate in 1864 was forty per thousand, higher than any other large city in the Western world. Prostitutes were said to be more numerous and more shameless than anywhere else on the continent, and the most influential woman in town was an abortionist, Madam Restell, who lived in a gorgeous, big townhouse on fashionable Fifth Avenue, ironically enough just across the street from where the new St. Patrick's Cathedral was being built.

Yet, for all its wealth, vanity, and squalor, New York City was the place to go, and there gathered the young, the talented, and the ambitious, not to mention the greedy and the unscrupulous. Yesterday's farmboy from upstate might be tomorrow's millionaire. Emporium czar A. T. Stewart had once been an ordinary shopkeeper living over the store, Cornelius Vanderbilt but a vulgar nobody from Staten Island, and Cyrus Field, the cable genius, came to town with but sixty cents in his jeans. As the undisputed center of the new America that emerged after the war, New York would attract the men of talent who, given time, would untangle the nasty snarls and delays in the streets that had been plaguing the population for a generation—men who would invent, promote, and construct the world's first elevated railroad.

Much of the city's transporation problem stemmed from the fixed boundaries of Manhattan—a long, narrow island hemmed in on west, south, and east by water. Before the completion of the Brooklyn Bridge in 1883 the only outlet for growth was toward the north, with the result that it had become a long, narrow city with its business section at the southern end and its homes at the northern extremity, an extremity that continued to recede as the business sector grew and pushed it farther away. Public travel from the middle-class uptown residential area to the downtown offices and stores was so pitifully slow, bumpy, and laborious by public horsecar or omnibus that the round trip might consume several hours. As the city grew gradually northward and uptown after the war, the daily trip became longer and more arduous, while tens of thousands hurried southward at the same hour in the morning and all rushed northward together again in the evening. Many families preferred to live in the suburbs across the river because they though commuting to the city by short ferry trips was easier and faster than the long trek to upper Manhattan. Some businessmen submitted to the long train trip each day from Westchester County to the north for similar reasons. Rapidly building up as a result were communities in southern Connecticut, eastern New Jersey, Long Island, and Staten Island—all to the loss of New York City.

But the working poor of the city did not have such choices to make in picking their neighborhood. Because of the barriers to city travel, movement about the metropolis was dreadfully difficult for them. Since stores and offices lay in the downtown area, employees had to live fairly close to their places of business in order to get to work. Lacking dependable transport, they could not spread out, but clung together like locusts at a rate of 25,000 to 300,000 to the square mile. Eighty-five percent of the city's inhabitants lived within two miles of the city center, then located at Union Square at Fourteenth Street. Only five percent lived farther than three miles away, and for most of the city's inhabitants the scale of distances was even smaller. Nearly half of the population lived in the narrow belt, one and one-quarter mile long, between Canal Street and Fourteenth Street. It seems ironic that New York City, the center of a worldwide communications and transportation network that moved men and goods across countries and oceans, was unable to break down the transit barriers within her own city limits.

Aware of the need to extend the city northward

in order to relieve the intense overcrowding that existed in the center, the city fathers after the war made every effort to open up what was called the New West Side of New York, an area between Central Park and the Hudson River. It was a pleasant area with good air and superb views of the river and the palisades on the Jersey shore, an excellent, spacious neighborhood for growing New York. Streets were extended there, parks laid out—Broadway and Riverside parks, Morningside Park, Edgecombe, and Colonial Park—along the rocky spine that extended up the island. Some new luxurious apartment buildings were planned in the area, such as the famous Dakota on Central Park West; but before the opening of the New West Side could be satisfactorily completed, one great problem had to be solved, that of local transportation. Without it upper New York would remain only a dreary wasteland of unpaved, ungraded streets crowded with squatters' shanties. Investors would not put up money to develop the area until some good means of travel could be found to get potential residents from their new homes to their jobs downtown in the city.

Not only was travel inefficient between the sparsely populated upper reaches of the city and downtown, but also it was nearly impossible to get around in the city center itself because of the wretched conditions of the streets and the horrid congestion caused by thousands of vehicles jamming the narrow, ancient streets. Horse-drawn conveyances of every description crowded together, for there were no limited access highways and no special truck routes, and no rail connections served the docks. Cargo and passengers were forced into the same struggling line of movement. Fortunately a city ordinance of 1867 excluded swine from built-up sections of town.

The streets themselves were badly paved, if at all. Wooden planks put down to keep wheels from becoming stuck in the mire split under the pressure of the constant heavy business traffic. Cobblestones were torn loose faster than they could be replaced, and in the region below Fourteenth Street, the sewers would often clog and break, spewing their awful reeking contents onto the roads and walkways. The primitive narrow pipes simply could not hold the water after a heavy rainstorm; thus such a major artery as Broadway, the city's grandest street, was turned into a sluggish stream of deep, brown mud after a rain.

And in dry weather when the hot sun dried out the mess, it took wing as dust and dirt that irritated the eye and covered the exposed fruits and vegetables on sidewalk stands. Potholes and ruts developed quickly all over town, causing New Yorkers with carriages or carts to thread their way through the streets with care lest their horse slip and break a leg. A horse in those days was a considerable investment, with a longer life expectancy than a modern motorcar.

The mayor of New York, John Hoffman, was constantly besieged by complaints about the filth that had accumulated in the public streets of the city. Filth, trash, and garbage endangered health and obstructed movement, sometimes totally blocking traffic. The paved roads of lower Manhattan were periodically cleaned by a private contractor, but his fee was not sufficient to pay for even weekly cleaning throughout the city and for the removal of snow and ice in the winter. Mayor Hoffman, testifying before an 1868 legislative committee, doubted that even larger appropriations would bring about notable improvement since the contractor was "not truly dedicated to cleanliness."

One of the greatest sources of filth in the streets was manure produced by the tens of thousands of horses that crowded the roadways. Animal power was the only power available for public and private transportation in New York City at that time, and a horse dropped over ten pounds of fecal material a day on the streets and periodically drenched the pavement with urine. We are told that in the busy streets the pavement was constantly wet with horse urine. Not only was all this offensive, but the feces carried the virus of tetanus so that any skin abrasion occurring on the streets might entail the risk of an absolutely horrible and fatal disease. Urination was so frequent that smooth pavements such as asphalt were not practicable; either dirt or cobblestones had to be provided to assure traction between a horse's hooves and the street. When New York's Metropolitan Board of Health was created in 1866 to combat the strong fears of a citywide cholera epidemic, one of its first enterprises was to remove 160,000 tons of horse manure from vacant lots throughout the city. New York smelled a little better. The social cost of horse traction had become so great and so offensive that the city government was under strong pressure to replace the animal-powered street railways with something cleaner, since those companies alone employed more than eleven thousand horses. However, until someone came up with a practicable solution for rapid transit, horses and their inevitable wastes would remain.

Even for all the mud, mire, mess, and manure the busy, noisy, cobblestone and dirt streets were painfully congested throughout the day. In the time just after the Civil War and just before the first elevated began running, horse-drawn streetcars and bobbing omnibuses fought their way through a medley of drays, hansom cabs, carriages of all descriptions, pushcarts, and vans. Traffic congestion, vehicular confusion, pedestrians in danger—these age-old problems plagued the New Yorker. The neighborhoods of greatest difficulty were downtown Broadway and the waterfront thoroughfares leading to the ferry slips on the Hudson and East Rivers. Wall Street was always a turmoil, because so much of its exchange business was conducted from the street. At Broadway near Trinity Church the clutter of the buses and carriages made travel a constant adventure of narrow escapes and near accidents. From Fulton Street north into the plaza opening up into what is now City Hall Park was a danger zone of the worst sort. Countless plans and proposals for the relief of transit jams and protection of travelers were advanced as early as 1840, though none were implemented.

Traffic was so terrible on lower Broadway in the commercial district that it was dangerous, if not impossible, for pedestrians to cross it during the day. Mr. Genin, an enterprising hat merchant occupying a busy spot across from St. Paul's Church, offered to erect a bridge across the street to allow shoppers to get across Broadway to his store. *Gleason's Pictorial Drawing Room Companion* heartily endorsed Genin's plan as well as the generosity of the "merchant price whose charity bids fare to supersede all others. At present this most dangerous crossing is in the highest degree hazardous for the ladies and children. By means of this ornamental structure they might pass over pleasantly and safely." Genin's bridge was never built, though another structure called the Loew Bridge, somewhat less ornamental in design, was put up on the spot in 1866, an attempt in a small way to provide an elevated sidewalk to relieve traffic dangers. Schemes for elevated sidewalks and pedestrian bridges above crowded streets would continue to intrigue the populace of New York for the next decade as the need to solve traffic congestion grew more urgent.

Even walking in teeming old New York was difficult as life overflowed the houses and shops into the streets. Merchants spread their wares across the sidewalks, and pedestrians and shoppers had to step gingerly into the roadway. Parked horse-drawn and hand trucks imposed their own rest on other vehicles, which were slowed to a crawl in the narrow channels. Housewives, hard-pressed to get about, stayed close to home in the course of their daily shopping. To meet their need, retailing was decentralized and retailers were mobilized. Peddlers trying to expand the size of their market roamed the city, their carts in many places transforming lanes of movement into stationary marketplaces and causing the streets leading into the buildings of the old public markets to be totally blocked.

The existing urban transit system in New York in the decade before the inception of the elevated railroad was hopelessly inadequate to handle the traffic. While the horsecar companies and omnibuses did a thriving business and continuous processions rolled over their lines, they were unable to transport the populace very satisfactorily because of their small capacity, slow rate of speed, and the wretched traffic obstacles in their way. It was a slow, bumpy ride just to reach the Hudson River Railroad station at Thirtieth Street, and the trip consumed more than forty-five minutes from the city center. And while steam-operated railroad trains would have provided possibilities for more rapid transit, New York, unlike most communities and cities in the mid-nineteenth century, did not welcome the intrusion of the railroads into their streets. The noise and smoke caused by the first crude engines created widespread indignation, frightening horses and residents alike with their deep-throated exhaust. After several locomotives had exploded in the streets, creating a panic and causing mobs of Irish to riot and tear up the tracks in the Bowery, the public outcry became so great that in 1850 the use of locomotives was prohibited below Fourteenth Street. Then as the population spread northward, the deadline in 1859 was set at Twenty-sixth Street, where a depot stood on the site of the old Madison Square Garden. Later the deadline for locomotives was fixed at Forty-second Street, which determined the location of the present

Grand Central Station. As a result of this restriction, though rapid transit was available in getting out of the city via railroads, the terminals were inconveniently located uptown, and only the lines of the New York Central System actually entered Manhattan. Because New York was a water-locked island, there was a gap in the transportation network to the outside world. Confined by the Hudson to the New Jersey shore, the other major trunk-line railroads to the west and south were required to ferry cargo and passengers across the river to Jersey City, the major point of transfer.

With the railroad excluded from their midst, New Yorkers had but few choices for public travel, all slow, all crowded, all unpleasant, and all animal-powered. They could ride in the omnibus (a sort of stagecoach affair with a large, cumbersome body) or a horsecar operated by one of the several street railway companies. The horsecars were larger coaches than the omnibuses and ran along iron track in the street. The omnibuses, legacies of a past era, were considered very old-fashioned in the 1860s; yet a dozen different companies continued to run hundreds of buses to many parts of the city as well as to neighboring villages. Broadway, where wealthy businessmen and residents resisted the intrusion of the newfangled street railway tracks, was served by these clumsy omnibuses until as late as 1906. The Broadway line started out at South Ferry, the busy ferry station serving Brooklyn and New Jersey, proceeded then over to Broadway and up along a portion of Ninth Avenue, eventually reaching Twenty-third Street, which at mid-century represented the northern extremity of the populated area of town. Other important north and south thoroughfares had omnibus lines similar to the one on Broadway.

The horsecars running in tracks were unquestionably an improvement in speed and dependability over the omnibuses, but their shortcomings were increasingly apparent as New York's urbanization progressed. Though the horsecar was reasonably satisfactory in small communities because it required little capital in track and was fairly cheap to run, in large cities such as New York where the climate was severe in the winter, even mild snows seriously interfered with horse travel and made the mode undependable. And when they were running in fair weather, horse traction was capable of only about four or six miles per hour, barely faster than walking. According to the *Scientific American*, it took a passenger "a dreary hour by horsecar to go from the Battery to Central Park." And in 1867 the *Evening Post* complained that workers had to spend more than four hours a day getting to and from their jobs. The need for the elevated railroad was great.

Nationally the horse railway was considered pretty much inefficient, but in New York there was as yet no alternative. The cost of maintaining a large stable of horses, usually a company's greatest investment, was a chief factor. The Third Avenue Street Railway alone had 1,700 horses, and the one running in Eighth Avenue reported 1,116. A car horse cost about $125 to $200 and was good for an average of four years of service before he had to be sold for less difficult duties; in the meantime he was likely to have lost most of his value. Then there was the expense of food, stabling, and shoeing. Mules were cheaper in first cost, stood heat better, and were somewhat cheaper to maintain, but they depreciated more rapidly than horses. Who wanted an old mule? Horsecar lines in New York City, where there were frequent stops and starts, needed two horses to a car and had to change them every few hours. Consequently, operators varied from owning from four to ten times as many horses as cars. Thus a major street railway necessarily had much of its investment in horses, an asset that depreciated rapidly and was subject to decimation from disease.

An event that has been used as a justification for building the elevated and one that stimulated the demand for small suburban steam engines was the Great Epizodic of 1872. During that summer there was an epidemic of a respiratory and lymphatic disease (something like an influenza) among horses in the eastern United States that within three weeks either killed or disabled 18,000 horses in New York City alone. The destruction of animals was so severe that gangs of immigrants were hired to haul horsecars through the streets manually. Thus though the shortcomings of animal-powered street railways were widely recognized before, the Great Epizodic, or the fear of its recurrence, caused people to think of the elevated as a rational alternative.

Considering the mid-Victorian sentiments of the period there were, of course, many humanitarian considerations made against the use of horses for

The "Horrors of the Horsecar" as depicted by Frank Leslie's Illustrated, *1865.* COURTESY PRINTS AND PHOTOS DIVISION, LIBRARY OF CONGRESS.

public traction. Contemporary accounts tell of the tragic facts of the misuse of horses; horses dropping dead from heat and overwork in the summers were not uncommon occurrences. Though most of the New York car companies established relief stations along their routes for the animals, public and private agencies had also erected 150 drinking troughs, and the Society for the Prevention of Cruelty to Animals had a force of ten men posted where they would be most effective for the prevention of abuse of the gentle beasts. Concerning the treatment of horses, Matthias Forney said, "Most of the people in large cities are now drawn from their business to their homes like pigs, and the animals that draw them are literally tortured to death, for overwork is torture of the worst kind."

Nor were the operators of the dozen or so horsecar companies thought to be very humanita-

rian toward their passengers. The New York press was continually berating them for their careless and callous attitude toward passenger comforts and public service. The cars were said to be filthy and smelly, lighted at night by one faint kerosene lamp and warmed in winter only by straw strewn on the floor.

In spite of the slowness and wretched condition the horsecars were usually crowded with passengers, because they were the best rides available. This popularity led to perhaps the most serious complaint about such travel—the overcrowding. It was not unusual for cars with a seating capacity of twenty-two to carry eighty-five or ninety. Passengers continued to climb aboard already full cars, packing themselves inside, on the roof, or on the front or rear platforms, evidently convinced that if one could establish a toehold, the horses would be equal to the load. Mark Twain denounced their slowness and accused city officials of winking at the overloading of streetcars. New York newspapers had a field day in the 1860s, describing what they called the "horrors of the horsecar." One paper grumbled that "modern martyrdom may succinctly be defined as riding in a New York bus." *Harpers Weekly* observed that the cars were "packed and jammed until there is scarcely room to breathe." The *New York Herald* of October 2, 1864, said, "Something more than street-cars and omnibuses is needed to supply the popular demand for city conveyance. . . .People are packed into them like sardines in a box with perspiration for oil . . . . the passengers are placed in rows down the middle, where they hang on by the straps, like smoke hams in a corner grocery."

The *Tribune* (Horace Greeley's newspaper) said on February 2, 1866: "The railroads and omnibuses have their uses, but we have reached the end of them. They are wedged for hours at night and morning with men, women, boys, and girls sitting, standing, hanging on . . . . they are unchangeably too slow . . . . Gentlemen of the legislature, give us both the Underground and the Aerial Railway."

The horrors of horsecar travel were very vivdly described by Simeon E. Church, an indefatigable worker for rapid transit and a master of Hogarthian prose: "Sixty people packed into a box so closely that no man can tell which legs are his own, and which his neighbor's, their heads buried in a dark concave above, breathing an atmosphere of fifty overfull or over-empty stomachs, mingled with a stench of rolling straw, the steam of reeking garments pervaded with the fumes of bad rum and worse tobacco or the pungent odors of undigested corn beef and cabbage."

In 1867 *Harpers* printed this caustic bit of doggerel called "Street-Car Salad":

Never full! Pack 'em in!
Move up, fat man, squeeze in, thin.
Trunks, Valises, Boxes, Bundles,
Fill up gaps as on she tumbles.
Market baskets without number,
Owners easy—nod in slumber.
Thirty seated, forty standing,
A dozen more on either landing.
Old man lifts the signal finger,
Car slack up—but not a linger—
He's jerked aboard by sleeve or shoulder,
Shoved inside to sweat and moulder.
Toes are trod on, hats are smashed,
Dresses soiled—hoop-skirts crashed.
Thieves are busy, bent on plunder,
Still we rattle on, like thunder.
Packed together, unwashed bodies,
Bathed in fumes of whiskey toddies;
Tobacco, garlic, cheese, and beer
Perfume the heated atmosphere.
Old boots, pipes, leather and tan,
And if in luck, a "Soap-Fat man."
Aren't this jolly? What a blessing!
A Street-Car Salad, with such a dressing.

In 1869 Junius Browne published a book called *The Great Metropolis: A Mirror of New York*, which contained a chapter entitled "Street Railways," filled with delicious sarcasm about the horsecar operators. "This city is for its sins occursed with at least twenty street-railways in the worst possible condition, running wherever one does not want to go, through the most repulsive quarters. The cars are odious and impossibly slow because of the congestion of traffic in the streets. Wagons bearing huge stones and ponderous machinery lie in wait for the cars, and break down across the track. Brick piles tumble at the precise hour one selects to go uptown, and cover the rails with impassable debris. Even trees blow down, and old women are seized with fits, and fall directly across the iron-bound way. Everybody and everything declare the railways nuisances, yet they endure and continue in the face of all opposition, and before the serious discountenance of the deities themselves. They declare they are merely anxious to accommodate the public and the only man who ever was accommodated by them died the next moment from the unexpectedness of the sensation. . . . The wonderful creature who renders street-railways impossible shall have a monument in Union Square higher than Washington's, and be represented on two horses. What is the father of his country compared to the mother of reform? The former was childless. The offspring of the latter will be blessed and unnumbered. Extinguish the street-railways, root and branch, and steam-cars, the greatest need of the metropolis, will supply their place."

When riding the street-railway, "one learns cynicism and feels suffocation in daily rides, through the sinuosities and odors of the filthiest streets. There is no monotony, some romance, much danger, and more disgust in the cars. Certain preparations are desirable, however, for the per-

Overcrowding on a Manhattan omnibus. Passengers continued to climb aboard already full cars, packing themselves inside or on the roof, evidently convinced that if they could establish a toehold the horses would be equal to the load. COURTESY THE SMITHSONIAN INSTITUTION.

formance. The regular passenger should lose his sense of smell; have the capacity to shut himself up like a patent umbrella; be able to hang on a platform by the lids of his eyes; hold drunken men and fat women on his lap, eight or nine at a time, without dissatisfaction or inconvenience, put weeping and screaming children in his waistcoat pocket . . . . be skilled as a pugilist and a crack-shot with a revolver. Those are the essentials for anything like resignation in the cars.

"A volume might be written on the drolleries and adventures of the railway victims. He who has ridden on the cars for a few years, and outlived it, is as interesting as a man who has been through the War, or thrice married, or half of his life a prisoner with the Indians. He bears a charmed life. He could jump over Niagara without disarranging his hair; or walk up to the bridal altar without trembling."

The need for rapid transit was undoubtedly grave. Without some solution the city would stop growing, and the overcrowding and congestion would grow worse. Without some fast and cheap means of travel between the business sector and the residential districts uptown conditions would grow impossible, and the result would drive families instead east and west across the two rivers and northward to Westchester County to find homes.

Eventually bending under the barrage of abuse hurtled at the street-railway operators by the newspapers, the State Legislature, long believed to be in the pay of the horsecar magnates, began to study proposals for rapid-transit systems. Public opinion over the disgraceful conditions was running so hot by this time that Albany had little other choice but to become involved. As a result the lawmakers were besieged by all sorts of rapid-transit ventures—all guaranteed to solve the city's transport malaise: subways, steam and pneumatic; elevated railways, cable, rope, pneumatic and sail; moving sidewalks, and iron tubes to be laid under the East River.

Clearly New York had to find some practical

20

solution to its transit miasma, meaning a plan that could operate at higher speeds than surface vehicles and be located physically either above or below ground, since the streets were far too congested already. And the solution would have to transport passengers in trains of cars rather than by single ones. Something like an underground railway or an elevated is justified only by the most extreme circumstances where there exists a large volume of traffic in a very congested area. These conditions were true in but a few cities in the world at this time in the mid-nineteenth century. Other than such metropolises as London or New York most urban centers were small enough in population or spread out enough geographically so that they did not suffer from impossible congestion. Consequently in most cities the omnibus and the horse railway were adequate to serve urban transit needs in the 1860s and 1870s. And by the 1880s the technology of cable and electric propulsion had improved surface transport so that again most cities were able to avoid the enormous capital expenditure and disruption of building high-capacity rapid-transit systems, either underground or elevated.

New York, however, because of its traffic, restricted land space, and physical situation, was forced to build either a subway or an elevated. Located on a narrow island between the Hudson and East rivers, New York City presented probably the most difficult situation in regard to internal traffic of any city in the country. Bounded on the west, south, and east by water, the only outlet for growth was to the north. As a result, early in the nineteenth century it became a long, narrow city with its business section at the southern end and its homes at the northern extremity—a boundary that constantly receded as the business sector grew and pushed it farther away. Also it should be pointed out that not only did New York have large numbers of people, but large numbers of middle-class people who could pay for public transportation, unlike Baghdad or Shanghai, for instance, where an expensive rapid-transit system would not have been economically possible.

The most obvious solution to the city's transit problem was to build an underground railway similar to the one that had been operating successfully in London since 1863. In fact a very well-organized project for a subway based on London's was launched in New York in 1864, and it received support from prominent investors as well as from the populace at large. However, when the backers sought official permission to begin construction, they encountered strong opposition that succeeded in preventing the State Legislature from granting the necessary authorization. So it was politics, not engineering, that postponed the building of an underground railway in New York for fifty years. Had the plan received the necessary rights from the Legislature, New York's first rapid-transit line would have been a subway. Instead in 1867 the Legislature chose the elevated railway, largely through legal circumstances but partly because it was cheaper and quicker to build than an underground.

The technology of the elevated railway was not really an innovation, but rather an adaptive technology. There was never anything very unusual about its engineering. With only minor modifications, such as burning hard coal for fuel, elevated-railway engineers took the already established steam railroad as a model and adapted it for city use—the locomotive, cars, and tracks. Then they took the already well-developed art of iron-bridge building and put them all together. Though the conventional steam railway was not good for surface travel in cities, it could go overground, or underground for that matter as in London, with little disturbance to the populace. Actually the potential for overground or elevated travel had been recognized as early as the 1830s by such men as Col. John Stevens and Robert Mills, who had urged the city fathers to consider such a system to relieve the streets of New York.

2 Early Schemes and Proposals for Elevated Railways

During the decades before and after the Civil War New Yorkers were entranced with the notion of rapid transit. While the citizenry hung from the straps of filthy horsecars, lurched perilously in clumsy, overcrowded omnibuses, and paid the ransom of piratical cabs, inventors were at work exploring the idea of putting an elevated railway above the streets of the city. Such a mode of travel was discussed widely and enthusiastically for forty years before it was actually tried in 1867, and some of the best-known engineers of the time worked on it.

The pages of the daily New York newspapers and the leading journals carried accounts of various weird and wonderful elevated-railway schemes, all promising the city relief from its transit malaise. Most of the articles included illustrations to show the reader how the inventor's structure would appear, and being proper Victo-

rians most of them were very careful to provide an ornamental superstructure—one that would add to the cityscape with graceful scrolls, Gothic arches, finials, balconies, and prettily capped Corinthian or Ionic columns. Some engineers were so proud of their artistry that they had colorful lithographs printed of their invention (such as did John Randal in 1848) and sold to the public for suitable framing. In spite of frequent, careful attention to architectural detail and fancy ironwork, there was often something fantastic about many of the schemes— rather as if some contraption from out of a Jules Verne novel had mistakenly crept in the night onto the pages of the *Scientific American*.

One bizarre roller-coaster elevated railway was put forward by Robert Cheeseborough, an engineer of some repute, who depended on gravity for power. His trains would always conveniently run downhill, thus eliminating the need for noisy

An elaborate scheme for rapid transit was proposed in 1872 by Dr. Rufus Gilbert. Inside tubes set atop huge Gothic arches spanning the street passengers would be shot around the city in projectilelike cars by air pressure. COURTESY *Scientific American,* APRIL, 13, 1872.

locomotives. Such a feat was made possible by a series of inclines, the carriage coasting from one elevation to another and a system of counterweights raising the carriage up to the next elevation after it had coasted to a halt; from thence it would glide off on its silent way once again. Another miscreant elevated invention was a scheme for a sailing train; all its pilot had to do was to pull a lever that raised a huge, silk sail from the roof, the slightest breeze setting the train in motion and sweeping it along the track. Approaching his station, the pilot would lower the sail and allow the train to coast to a stop. Here was a simple, cheap idea, but unfortunately the inventor made no provision for running in a headwind or on a still day.

Since weight was naturally a consideration in elevated travel, another inventor devised a picturesque vehicle carrying a long, dirigiblelike bag of hydrogen on top to lend bouyancy. Then there was the bicycle locomotive, an incredibly slender train that had three wheels on each car in a single line and ran on a single rail. A pulley attached to a

cable overhead prevented the train from falling off, its inventor hoped, while it soared gracefully over the congested New York streets, descending to earth occasionally to pick up passengers.

There was also an idea called the viaduct plan, to build a continuous row of high stone houses from the Battery at the lower tip of Manhattan all the way up to Harlem at the top of the island. Small steam locomotives would carry cars across the rooftops of specially constructed buildings and over stone bridges across the streets. This plan was especially appealing to the infamous Boss Tweed and his Tammany ring because it offered the potential for such enormous graft. Several other ingenius proposals were made for elevated trains running through a network of pneumatic tubes through which passengers tucked inside metal cylinders would be shuttled about the city like charge slips in a department store. A lot of creative imagination was at large in those days.

Although some of these early schemes may seem fanciful, even foolish, in retrospect, the growth of elevated railroad technology is, of course, a serious part of nineteenth-century transportation history, whose antecedents go back to the early years of the century when they were called "aerial railways." The first man to propose a steam-operated elevated railway was Colonel John Stevens of Hoboken, a member of a wealthy and politically powerful New Jersey family. In the surprisingly early year of 1812 Stevens suggested a wooden trestle railway be built upstate between Lake Erie and the capital in Albany. According to Stevens the wooden elevated could be constructed quickly and would be cheaper than a road or canal, since the elevation would eliminate expensive grading and ballast. The railroad was to be powered by a small steam locomotive. It is important to remember that in 1812 Stevens's idea was particularly striking, because it was only the dawn of the age of steam and the only steam conveyances in America were boats. The few locomotives that existed at that time were in England. However, the locomotive was becoming a commercial reality, and men of pioneering vision such as Colonel Stevens were beginning to think about such locomotion.

When Stevens approached the Commissioners for the Improvement of Inland Navagation with his idea, he was received with understandable coolness because his plan was so far ahead of its

One of the most ambitious plans for an elevated railway in New York was the patent of Alfred Speer. He proposed to construct an endless, continuously moving platform on which passengers could stand, walk, or ride in enclosed drawing rooms.

time. Nowhere in America had anyone attempted to run a locomotive on a line of iron rails—elevated or otherwise. Though the idea caught the fancy of men who were seeking a solution to the transylvanian transit dilemma, the commissioners never took up his prophetic idea, perhaps because it was too visionary but more likely because they were too caught up in canal projects. Canal fever was sweeping the country in 1812.

The first man to suggest a steam-operated elevated railway in New York City was also John Stevens, a pioneer in the realm of transportation until his death. He announced his idea for such a road to the City Corporation in 1830, to commence at the Battery and run up through Greenwich Street to some point above the State Prison, then to turn and rise in a bridge across the Hudson River to Hoboken on the Jersey shore, where it would connect with a canal. The railroad was to be carried on a structure of a double row of wooden posts over each side of the street and elevated ten or twelve feet above the pavement to admit carriages to pass underneath freely. In addition to carrying passengers the colonel planned for his el to transport goods, produce, water (in those days before Croton Water New York City had a perennial problem with the water supply), and coal (the city would welcome a plentiful supply of cheap Schuylkill coal). Again he had to battle the powerful canal lobbyists in Albany, where the idea eventually died. In his attempts to secure state approval for the railroad, Stevens often complained of the formidable coalition of lobbies who opposed him: canal companies, dredging companies, steamboat lines, stage lines, and turnpike companies. He said that even the farmers selling feed to stagecoach horses and the innkeepers along the stage routes lobbied at the state legislature against him. Unfortunately Stevens died in 1838 before his idea of elevated railroads became more generally accepted. It is interesting to note, however, that Stevens did erect an elevated pleasure railway at his Hoboken amusement park called Elysian Fields. The diminutive el was a favorite with New Yorkers for many years.

Perhaps inspired by Stevens's idea, other men began to propose variations on the same elevated railroad theme. Always the cheapness of construction was a central argument for the stilt railway. In 1826 the prominent architect Robert Mills published plans for an elevated monorail to connect Washington and New Orleans. And about the same time the Delaware and Hudson Railroad was built in eastern Pennsylvania, largely on an elevated structure because of the rolling terrain it traversed. Far to the south another trestle railway appeared, the creation of Horatio Allen, chief engineer of the South Carolina Railroad, who believed the elevated plan particularly well suited for the cotton states where good lumber was in plentiful supply. He also convinced backers of the road in Charleston that the trestle plan would be safe from floods and require no expensive fencing because man and beast could pass freely under the structure. Between 1830 and 1833 Allen's bizarre stilt railway was built across the state, but just months after the opening the scheme proved to be a dreadful mistake. The presumably hearty yellow pine decayed rapidly, and the structure sagged and rotted quickly in the subtropic climate near the coast. The rickety tracks became very dangerous, since what might have been a minor derailment on a ground level line could turn into disaster on an elevated. Realizing his folly, Allen agreed to rebuild the line as a conventional railroad, and so seventy-five miles of the jerry-build structure was filled in with an earthen embankment.

The failure of the South Carolina Railroad did

Although he never won support for his elevated in New York City, Colonel John Stevens did construct an elevated pleasure railway at his Hoboken amusement park, the "Elysian Fields," just across the Hudson River. The little el was popular with New Yorkers during the 1830s and 1840s. COURTESY THE SMITHSONIAN INSTITUTION.

*The Greenwich Street el set to music: the popular
interest in elevated rail travel after the Civil War is
reflected in this sheet-music cover, "Rapid Transit
Galop."* COURTESY THE SMITHSONIAN INSTITUTION.

not discourage others from trying the elevated railway idea. At least three others were built about the same time: the Syracuse and Utica, the New York and Erie, and the Ohio railroads. The Ohio Railroad was one of the more amibitous of several Western railroads projected in the Jackson era. In 1839 the company had completed a portion of their line between the towns of Fremont and Maumee in the northwestern part of the state when they fell upon hard times following the panic of 1837 and failed. Somewhat later the records tell of a novel elevated railway built by a logging company through a forest in Sonoma County, California, where trees were sawed off level with a cross-cut saw and the ties fastened upon the stumps. Since some of the trees were huge redwoods, there was firm support for the heavy, timber-loaded trains. In spite of a few isolated examples, aerial railways before the 1850s were thought to be curiosities and exceptions to the norm.

Then in the 1850s a serious epidemic of elevated fever struck in New York City. A great swell of popular interest developed in the idea of rapid transit in general and in elevated rapid transit in particular. No doubt realizing that the appalling traffic congestion and growing population would only make the future of the city more difficult, New Yorkers of all sorts turned their attention to elevated travel as a cure for the city's ills. Countless proposals were put forward and patents declared. The pages of the *Times, Tribune,* and *Herald,* the leading weeklies such as *Harpers, Leslies,* and *Gleasons,* and the technical journals, especially the prestigious *Scientific American,* reported on the various elevated contraptions submitted by a host of imaginative inventors. In a short time the need for the elevated became a popular cry. To meet the public craving transit companies sprang up in second-floor offices along Broadway, each with a patent, a prospectus, and a collection of testimonials from experts who had studied the plan and pledged it to be viable, safe, decorative, and ideally suited to New York's needs. All that the inventors lacked was capital, and the public was invited to invest and share in what one was assured would be handsome profits. Many did invest. But many more were bitterly opposed to the elevated railways. Furious protests from shopkeepers and residents along contemplated routes resulted in a storm of letters to the editors of the city newspapers, claiming the high roads would shut off sunlight and turn their streets into dark and dangerous canyons. And if locomotives would pull the elevated trains, they would spew hot cinders on people and houses; if they were horse-drawn, they would drop the obvious into the streets below.

Amid the protests the elevated fever raged on. Several possible reasons account for the interest in elevated travel in the 1850s, in addition, of course, to the pressing need for it. At this time New York was going through the throes of its first big street railway boom, which naturally aroused a good deal of popular interest in urban transit. Men came forward with schemes they felt were superior to the horsecar; many of these counterproposals were for mechanically operated railways rather than horsepower, and many of the inventions ran on tracks elevated above the street level while others ran underground. The idea of an elevated railroad had much merit, because it was relatively cheap and relatively fast to build. But because the city was so wealthy, many serious engineers believed that the underground with its solid tracks, superior speed, and invisibility was a better choice. Compared to the subway the elevated was visually rather ugly and a nuisance to street traffic. Thus various plans for subways appeared, but the fears of the population of riding underground and the consequential diseases that were thought to result from the darkness, gases, agues, and plagues were enough to discourage most promoters.

Popular interest in elevated travel was stimulated by New York's Crystal Palace Exhibition of 1853 (held at a site on Sixth Avenue and Forty-second Street now called Bryant Park), which aroused great excitement in commerce and industry. Moreover, it was at this same time that London's subway was under construction, leading many New Yorkers to think about their own need for better urban transportation.

Although some of the projected elevated railways appear hopelessly impracticable, many had a technical sophistication and an alluring, imaginative vision about them. The sheer ambition of some of the inventions is almost dazzling. Here are some of the more viable proposals for elevated for New York, gleaned from the brittle pages of the *Scientific American.* T. M. Brennan, a mechanical engineer from Nashville, Tennessee, proposed to propel cars on an elevated structure by condensed

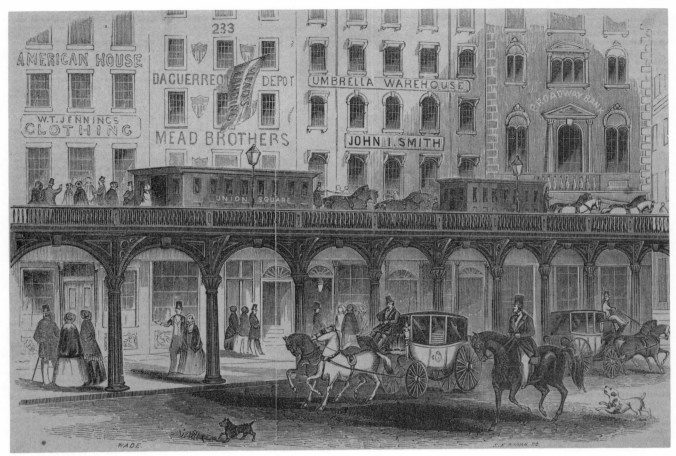

J. B. Wickersham suggested an elevated terrace along Broadway with a horse railway on the outside and a pedestrian promenade on the inside next to the building line. COURTESY Gleason's Pictorial Drawing Room Companion, APRIL 1, 1854.

air, an idea he believed well suited to city use because of the complete absence of pollution from noise and smoke. J. E. Holmes, superintendent of the Machine Department at the Crystal Palace agreed, saying that, "Sooner or later according to the length of the reign of fogyism, there will be an elevated railroad up Broadway, the cars of which will be propelled by condensed air."

Another idea was suggested by James Swett, an inventor from Pittsburgh, who planned to run a steam locomotive carrying a passenger car slung beneath it. The rails would be erected on arms branching like a lyre from strong pillars placed near the curb so as not to interfere with street traffic. Swett planned to fuel his engine with coke, not wood, which he thought "might set fire by a stray spark to one of Stewart's bales of fine French muslins." In a 1853 letter to the *Scientific Ameri-*

can Robert Mills, architect of the Capitol in Washington City, disagreed with a recent proposal to build an underground railway beneath Broadway similar to the one in London, favoring instead "passing our railroad trains through cities by elevating them above the heads of pedestrians." His principal objections to the "tunnels," as he called them, were the great expense, inconvenience, and interruption to business from the lengthy excavation, plus the potential injury to passengers' health from deadly vapors while traveling underground. Mills's elevated was similar to the one Charles Harvey built in the late 1860s; however Mills planned to run horse-powered omnibuses on his, not steam locomotives. The *Scientific American* did not very much cotton to the idea of the elevated omnibus, but did rather like his use of an auxiliary second-story sidewalk to relieve street-level congestion.

J. B. Wickersham, a well-known New York iron-railing manufacturer of the day, presented yet another plan for an elevated terrace along Broadway with a horse railway on the outside and a pedestrian promenade on the inside next to the stone fronts. The raised promenade over the sidewalk would be supported by rows of graceful, classical beams and ornamented, of course, by pretty iron railings. To deaden the noise overhead, Wickersham thoughtfully planned to lay the rails on India rubber sills to absorb the vibrations. According the *Gleason's Pictorial Drawing Room Companion* of April 1, 1854, the design was "entirely feasible, and finds many ardent supporters among the people of the metropolis, and the frequenters of that great thoroughfare."

Since many Broadway businessmen were opposed to the use of their sidewalks for the elevated, Charles Mettam obliged by putting his high road down the middle of the broad street on iron pillars and suspending the track on arms branching out on either side. About the same time, William Hemstreet had a grandiose scheme for an elevated that would have entirely covered Broadway with an arcade, on top of which would travel two railways propelled by cables and stationary steam engines. Passengers could board from balconies from the second floor of shops, museums, and tenements along the route so that New Yorkers would no more have to go "out into a filthy, dangerous street for a bounding, thundering omnibus." Furthermore, the plan would save many thousands of dollars in street cleaning by catching the rain and snow that would run off through the hollow iron pillars. And since the top of the glass-roofed arcade would be transparent, it would not darken the street as had many other projects.

During the nineteenth century various new forms of traction were being proposed, including a system known as the atmospheric railway. Atmospheric propulsion enjoyed a real vogue in the 1840s, and four such railways were actually built— one in Ireland, two in England, and one in France. I. K. Brunel's broad gauge, which ran twenty-one miles from Exeter to Newton Abbot in 1847–48, was the longest and most ambitious of the atmospheric systems. The St. Germain line outside Paris ran for the longest period of time, from 1847 to 1860.

To harness the power of the atmosphere, stationary steam engines were built every two or three miles along the track, beneath which was buried a cast-iron tube. Working air pumps, the engines exhausted the air from the tube, creating a vacuum. A piston was attached to the railway cars overhead, and the power of the atmosphere behind the piston alone drove the train. After the apparent success of Brunel's South Devon Company such railways were proposed in great number all over the British Isles as well as in Belgium, Germany, Austria, Italy, the United States, and the West Indies. And for a time atmospherically powered trains were thought to be the transport of the future, because they were silent, smokeless, clean, and relatively safe. Though seemingly practical, atmospheric-powered trains needed too many details to be perfected, and the idea shortly failed as a viable system. Such railways were expensive to build for one, and in many ways they were operationally inconvenient, hard to stop with accuracy, and with no way to control train speed by the engineer. Although interest in long-distance atmospheric railways languished after the late 1840s, the idea of using air pressure for urban-transit schemes continued through the rest of the nineteenth century. In fact air pressure was widely adopted in American and Western European cities in the 1880s and 1890s to transport messages, parcels, and mail.

Among the adherents of atmospheric power, or pneumatic propulsion as it was often called in America, was Dr. Rufus Gilbert, who designed an elevated railroad thought to be especially tasteful and ornamental, calling as it did for huge Gothic iron arches spanning the street from curb to curb. Though a bit late with his idea, Dr. Gilbert proposed atmospheric propulsion to move his elevated cars. Inside his gorgeous arches would run a double set of atmospheric tubes eight or nine feet in diameter through which cars propelled by air pressure would carry passengers about the city. Stations about a mile apart would be fitted with elevators to take passengers up to their cars so that there was no trudging up and down steep staircases. This impossibly impractical scheme actually got as far as the State Legislature before it was defeated by property owners. In those days apparently everyone wanted rapid transit, but one group or another always opposed any concrete plan when it came up for a vote. The indefatigable Dr. Gilbert, however, did not give up easily. We shall hear more from him in subsequent chapters, for even though his patent elevated highway was not constructed nor his method of operation adopted, the franchise that he eventually received

The Meigs Elevated Railway offered a raised engineer's turret and an observation deck on the locomotive. Passengers rode in streamlined, sausage-shaped cars, which ran on an elevated monorail track.

was used later to construct the Sixth and Second avenue lines on Manhattan.

One of the most colossally ambitious of all el schemes was proposed by Alfred Speer of Passaic, New Jersey, whose "quickest and cheapest" plan was presented in great detail in his pamphlet entitled *Treatise on City Travel with a True Solution of Rapid Transit*. Complete with graphs, drawings, charts, figures of costs, and endorsements from prominent engineers, the booklet explains how the inventor would build an endless, continuously moving platform on which passengers could stand, walk, or ride in enclosed drawing rooms at a speed of about twenty miles an hour. So that the platform would never have to stop, transfer to and from the contraption would be made by open shuttle cars, which would stop at each street to receive or discharge riders.

These are but a few of the scores of proposals for early elevated travel that turned up in the public records. Actually there were many more. The patent office received a hundred and twenty-five plans between 1825 and 1899. In 1867 alone more than forty drawings for elevated roads were submitted to the New York State Legislature. So many schemes poured into the offices of the *Scientific American* that at one point the editors eventually became rather bored by the whole idea, saying, "The subject is to us become somewhat *dry*." And they refused to publish any more of them.

Although few of the schemes ever got further than the drawingboard, the patents do indicate how the minds of inventors and engineers were working about the middle of the nineteenth century and even before and what their conceptions were of practical intraurban rapid transit. Of all the schemes only one actually came to fruition, that of Charles T. Harvey, whose experimental cable-operated elevated railroad went up in lower Manhattan in 1867.

30

3 The First El: Charles Harvey's Cable Line on Greenwich Street

While the shameful travel conditions that existed in New York during the middle 1860s demanded immediate relief, the public authorities did precious little to provide a solution. Despite the number of imaginative engineers struggling to prove the success of inventions intended to give the city speedy, efficient urban transportation, their efforts were blocked in both the city and state governments, who refused to take action on the urgent question. The reluctance of the politicians, complicated by so many rival interests and the opposition of the city's street railway and omnibus companies, led to a long and involved legal quarrel that continued for several years, holding up the development of an effective transit system and leaving the public the principal loser.

To grasp fully the recalcitrance of the State Legislature, one would do well to study how that group, the governor as well as the senators of New York State, consistently ignored many viable proposals for elevated and underground railways for New York City during the decades of 1850 and 1860. In particular, one plan for a subway to be built up Broadway from the Battery to Central Park deserved more generous consideration than it received. Sponsor of the subway and promoter of the Metropolitan Railway Company of 1864 was Hugh B. Willson, a railroad man from Michigan, who had lived in London during the construction of the London Underground. So impressed was he with its merits that he conceived of the idea of building the same kind of road in New York. As a result, on his return ot the city he sought the support of prominent men to finance his venture, securing backing to the sum of five million, a handsome figure for those days. With such impres-

Harvey's cable was carried by small, four-wheeled trucks called travelers, which were attached to the cable every 150 feet. A horn at the top projected up through a slot in the tunnel box and made contact with a grip on the car.

sive backing the subway got a good deal of attention from the press and the public, who seemed to like the idea of an underground railroad. On April 11, 1864, the *New York Times* said editorially that the project promised to be an immense boon to the city.

In the Legislature, however, the Willson subway was not received with such enthusiasm. Horsecar and omnibus companies had enormous influence with the politicians, and consequently the underground plan was plainly ignored by the Railroad Committee of the Senate, who simply put it off for consideration until the next year. Incensed by the high-handed action in Albany, the public became aroused, petitions were circulated about the city asking for reconsideration of the bill, and the press denounced the Legislature and called for reform measures. When Willson's bill came up again the next year before the lawmakers, the extraordinarily well-organized and well-engineered subway plan was passed, but to everyone's surprise it was vetoed by Governor Fenton. Undaunted by the two rejections, Willson and his aides continued to improve their plans and gather support wherever they could. When the fight in the Legislature opened the following session, Willson was confident that he could overcome the objections of the governor and win his charter, but he was put off once more, this time by Alfred Craven, influential chief engineer of the Croton Aqueduct Board (New York City's water system), who objected because he thought the subway might interfere with underground water pipes. This final obstacle by Craven proved to be the end of Willson and his Metropolitan Railway. Although first in the field with a well-organized plan, his project was killed by the dictates of the powerful horsecar corporations whose greed had for years kept the priceless franchise on Broadway unshared. Through years of quarreling by the politicians the struggle for rapid transportation was still at a stalemate.

With the death of the subway project the future of congested New York appeared grim, for not only had the legislators killed the Willson plan, but several other plans for elevated railways as well. Not all of the state senators, however, were in the pay of the horsecar lobby. There were a number of friends of rapid transit in the body—men who, like the inventors and their supporters, were frustrated by the years of delay and stalling. One such man was Senator Samuel Ruggles, who, after the defeat of the Willson bill, requested that if Governor Fenton would not support any of the proposals he should appoint a special commission to study the transit problems of the city and work out a plan for efficient travel for the city, coming up with the best system—be it elevated or subway—as well as the most advantageous routes through town. Ruggles's action in April 1866, together with the passage of a bill that attracted little attention at the time, eventually gave New York its first rapid-transit system, the elevated railroad. This apparently insignificant bill was passed to provide for the formation of companies to operate elevated railways by cables powered by stationary steam engines. State engineers had studied such a proposal and found it potentially practicable and worthy of an act from the Legislature to allow companies to form and to interest capitalists to begin actual construction. The bill was passed in the interests of Charles T. Harvey, the so-called father of the elevated railway. In the clash of rival underground and elevated schemes, the bill had slipped through without attracting much notice and without its importance being generally recognized by the legislators in the pay of the New York horsecar lobby. It nevertheless paved the way for the first urban elevated railroad in the world: the West Side and Yonkers Patent Railway.

The select Senate Committee on Rapid Transit requested by Senator Ruggles was quickly formed, consisting of George Andrews, N. R. Low, Charles G. Cornell, all of the Senate, plus John Hoffman, mayor of New York, and Alfred Craven, engineer of the Croton Board. One of their first steps was to advertise for plans, allowing four months' time for their preparation and submission. By October a great many drawings had been submitted, and in November the commission met to begin their consideration and, hopefully, their selection of a plan that would have the desired elements of speed, safety, cheapness, and quickness of construction. Among the plans brought forth were those from A. P. Robinson for an underground railway; from O. Vandenburgh for an underground to run in beds of gutta percha to minimize noise; from M. O. Davidson for a pneumatic subway system; from S. B. Nowlan for an arcade and basement plan; from John Schuyler for a depressed railway scheme; from James Swain for a combined underground, surface, and elevated road; from Gouverneur Morris and Isaac Colman for a very expensive viaduct-plan elevated railroad; and from Charles Harvey for an elevated railroad powered by an endless cable.

After careful consideration the commission unanimously recommended the underground railroad as the best and quickest type of transit for New York; however, they stated that Harvey's plan for an elevated deserved consideration because it was carefully prepared, was most free from engineering difficulties, and because of the use of cable traction was free from the pollution problems of steam locomotives. Thus while the committee recommended the adoption of a subway system as most suitable, it did not in its report choose any one of the several plans submitted for that kind of a road as best. Apparently, since the body was unable to agree on any specific project, they decided to let the rival companies fight it out before the Legislature for the coveted franchise.

Encouraged by the report, Harvey and his associates incorporated in July 1866 the West Side and Yonkers Patent Railway, proposing to build a twenty-five-mile-long elevated railroad from the Battery up Greenwich Street and Ninth Avenue to Kingsbridge and Yonkers in the northern suburbs. Another line was planned on the East Side, mainly up Third Avenue to the Harlem River and on to New Rochelle. The lines were to be constructed under Harvey's patents, operated on a special elevated structure of his design by endless cables, and powered by stationary steam engines placed in street vaults located at intervals about 1,500 feet apart along the route. Harvey estimated the cost of his railroad at from $250,000 to $500,000 per mile, or about a third the cost of the only other elevated scheme submitted to the committee, that of Gouverneur Morris, who wanted to build an elevated through city blocks, requiring a swath of land up through the city fifty feet wide as a right-of-way. Cost of constructing the Morris elevated was estimated at $1,165,000 per mile, not including the expense of the real estate for the right-of-way.

Early in 1867 Harvey applied to the Senate for permission to build an experimental half-mile section of his elevated in Greenwich Street on the lower West Side of New York. His petition read in part: "The undersigned Memorialists, being citizens of and property-holders in the city of New York, respectfully beg leave to represent that they are impressed with the belief that the greatest public want of the city is a new method of transit between parts on Manhattan Island and the Northern suburban villages. . . . No locomotives are used on the proposed railway, as the motive-power is confined to a series of noiseless, endless ropes, which is driven by engines made stationary beneath the street pavement, consequently neither smoke, cinders, oil, or noise can be offensive to the ordinary users of the public streets. . . . If it does not answer the purpose, it will be for the interests of your memorialists to abandon the project, and remove the trial line at their own expense."

Harvey's petition was received favorably by the Legislature, who passed a resolution permitting him to use the public streets for his trestle with the stipulation that he complete his work within one year. If after inspection by the state railroad commissioners his patent railway proved successful, he would be permitted to extend the line up Ninth Avenue as far as Yonkers. The act was promptly approved by Governor Fenton, thus making way for New York's first pilot elevated railway. It is interesting to note that at the same session the Legislature granted a charter for John Roebling to build his Brooklyn Bridge across the East River. The two great works were needed for nearly the same reason, according to the *Times,* to ease the city's most serious problems, crime and overcrowding.

Fares for the el were fixed by the Legislature at five cents for any distance less than two miles and one cent for each additional mile. After five years the company was allowed to set a uniform rate of ten cents for a ride. In return for the franchise the company was to pay five percent of its net income to the city for use of the public streets. In an

The original cable car was lowslung to give passengers a feeling of stability. It stops here at the Twenty-ninth Street station on Ninth Avenue. The horse pulling the omnibus seems quite undisturbed by the commotion.

attempt to make the railroad as unobtrusive as possible the charter also set careful controls on Harvey's motive power (limiting him to the use of engines put invisibly under the surface of the street) and on the construction of the elevated structure itself (dictating the strength and height, and preventing undue blocking of businesses and houses along the route).

In securing the valuable franchise for his invention, Charles Harvey deserves considerable recognition, for the Legislature was in those years not easily moved by unproved transit schemes. But since Harvey was a fairly well-known engineer with some influence in the state, not a callow dreamer, his reputation may have swayed the lawmakers. Like a good many other successful men in New York City after the war, Harvey was a Connecticut Yankee, born in 1829 in Thompsonville, a village adjoining Hartford, where his father, the Rev. Mr. Joseph Harvey had for more than thirty years been pastor of the Presbyterian Church. As a young man Harvey's first responsible position was as a salesman for the Fairbanks Scales Company; after a few years he was appointed managing agent for the "western states," which meant the region from Pennsylvania to the Mississippi River. In 1852 he was made general agent for the Fairbanks Company at Sault Ste. Marie, Michigan, a typical frontier town of the period. The "Soo" was at that time beginning to take on added importance, since Congress had authorized the construction of a canal to join Lake Superior and Lake Huron. Because the constitution of Michigan did not provide authority

for incorporating a company to construct the canals, the New York based Saint Mary's Falls Canal Company was chartered with a capital stock of $1,000,000.

Charles Harvey was appointed resident manager of the canal project, which was in part financed by the Fairbanks Scale Company, later rising to chief engineer because of the high quality of his work. When ground was broken for the "Soo" canal in 1853 it was the largest public works project in the country, employing at peak times between two and three thousand men. Among the many problems confronting Harvey was the transportation of materials. No railroad ran to the area, and Detroit, the nearest manufacturing center and chief source of supplies, was entirely cut off during the winter months when lake traffic ceased. The contract required the completion of the canal by June 1855, but in the winter of 1854–55 an unexpected difficulty arose that tested Harvey's engineering ingenuity. Through an error in the official survey of the canal site, a sandbar turned out to be a granite reef. Because there was no way of transporting the necessary machinery to break up the rock formation during the winter, and with the time element of the greatest importance, Harvey saved the day by devising an improvised tool consisting of a drop hammer, fashioned from the propeller shaft of an old steamer driven by a hoisting engine. By means of the improvised machine the granite ledge was broken up and removed, thus assuring completion of the project within the specified time.

Grateful for his resourcefulness and skill, the president of the canal building company, Erastus Corning, passed a resolution of commendation for the chief engineer. The document, suitably framed in the Victorian manner, was presented to Harvey, stating that, "The discharge of duties of his position under great difficulties entitles him to our thanks." Later on his service to the canal builders would prove helpful to Harvey, for in 1871, when his elevated railway was under fire from Boss Tweed, he was to call on the politically influential Erastus Corning, who never forgot Harvey's service to him and who came to the engineer's aid at a moment of crisis.

After completing the canal project Harvey became interested in transportation. In 1855 he went to Washington to lobby for federal support of railways in the Northwest and was successful in securing large land grants for the railroads then spreading their operation into Michigan and adjoining states. Later he was associated with the construction of the first railroad to reach Lake Superior, now a division of the Chicago and Northwestern Railway. In 1858 he worked as engineer erecting a furnace for the Northern Iron Company, of which he was managing director. In fact the village that sprang up around the ironworks on the south shore of Lake Superior near Marquette, in the upper peninsula of Michigan, was named Harvey after him.

Having earned a sizable fortune Harvey moved East with his wife, settling in the fashionable Hudson River community of Tarrytown, New York. Not long after his return East he began to turn his attention to some solution to New York's intolerable transit malaise. Even before the Rapid Transit Commission advertised for plans for a practical system for the city, Harvey had begun to work out the details on his idea for an elevated railway powered by endless cables. Mindful of the stiff competition his invention would face, he prepared elaborate pen-and-ink drawings and detailed models to illustrate his work. One, a working scale model with a circular trestle ten feet in diameter, was equipped with live steam from a tiny copper boiler, which drove two diminutive cables connected to cars that ran gaily around the track. This and two full-size sections of his structure Harvey took to Albany and had set up in the plush lobby of the American Hotel, near the state capital building. The politicians whom he invited to inspect his invention were apparently charmed by the Lilliputian railway, for on the day after his inspection party Harvey's elevated railway bill passed the Legislature with an overwhelming majority.

Harvey lost no time in getting started with the task of building his elevated. Within slightly more than two months after the official go-ahead from Albany he began the preliminary field work of surveying the right-of-way for the first quarter-mile section in Greenwich Street from the Battery to Morris Street in the area just off the present financial district. Excavations for the subterranean steam engines that would power the cable were well underway by early July, and by October the first columns were being put up. Then on December 7, 1867, on the same day that Charles Dickens arrived in New York for his second visit to give a series of readings for the approaching Christmas season, Charles Harvey demonstrated the safety of his elevated to his backers and directors by taking a short ride. Dressed nattily in a black frock coat and high, beaver hat and perched proudly in a small handcar, the inventor sailed smoothly along the second-story level of lower Greenwich Street, his black beard fluttering gaily in the chill morning breeze. Below him his delighted directors looked on, smugly assured of their investment and somewhat dizzy with the rich future in store for them. Fortunately, a photographer was on hand to record the momentous ride, capturing one of the rarest scenes in New York history. As the picture illustrates, the event did not draw large crowds. Only a few invited guests were present that early Sunday morning; the official test trip for the railroad commission was yet six months away. This demonstration was for friends of the West Side and Yonkers Patent Railway, who as it turned out were so pleased by what they saw that they authorized Harvey to go ahead with the second quarter mile of track to Cortland Street. Moreover, they asked him to order a proper railway car so that next time they could all take a ride.

Behind Harvey and the world's first rapid-transit urban elevated were a number of prominent businessmen and financiers who gave him encouragement, moral support, and money. Directors and major stockholders of the original company were William Appleton, the publisher; Chauncey Vibbard, one-time congressman from New York, director of the United States government's Military Railroad during the Civil War, and organizer of the vast New York Central Railroad system; Walter Gurnee; Frank Work; Samuel Pettingill; Moses Hoppock; John Perkins, and William E. Dodge, wealthy philanthropist, owner of the Phelps Dodge metals corporation, and backer of several railroad affairs. So principled was Dodge in business matters that he severed his connections with the Erie and Central Railroad of New Jersey railroads when in opposition to his objection they continued to run Sunday trains. Other friends of the Harvey venture were David Crawford and Peter Cooper, perhaps the best-loved and most respected man of his time in New York City, possessor of a large fortune based on a profitable glue factory and iron foundry and founder of the Cooper Union, a free school for the

The inventor Charles Harvey demonstrated the safety of his cable-operated elevated railway on Sunday morning, December 7, 1867, by taking a ride down Greenwich Street in a small handcar. His company directors look on from in front of Edwin Richards's warehouse at the corner of Morris Street.

working men and women of the city.

With the solid support of his backers Harvey pushed ahead with his construction, ever mindful of the completion date of July 1868 set by the governor. The date was very important because he had to comply with the schedule in order to receive approval from the state to lay more track northward, so necessary to put the road on a paying basis. And so, block by painful block the thirty-foot-high, one-legged elevated superstructure crept up Greenwich Street, past the second-floor windows of banks, hotels, saloons, stores, warehouses, and dwellings. Since the columns supporting the track were set in the sidewalk just inches inside the curb, the cars would run very close to the building facades along the route; in fact it would be nearly possible for someone living in a second-floor room to reach out and touch a passing cable car.

As the trestle crawled steadily forward, curious crowds gathered to watch the workmen. They expressed almost universal skepticism over the whole idea, jeering that the flimsy stilts would never be able to stand the strain of a loaded car but would surely collapse into the street. Harvey, however, was an experienced engineer and knew better than to listen to the jibes of the street people, nor did he have the time to ponder failure.

During daylight hours he was busy supervising construction of the iron columns and track, and at night he crawled into one of his vaults beneath the sidewalk and worked at perfecting the engine-cable transmission. Although the columns may have looked weak to the average pedestrian, they were especially designed by Harvey and quite strong, being fabricated of wrought iron eighteen inches in diameter at the bottom. He had found a patented column at Phoenixville, Pennsylvania, which he adapted for the elevated posts by riveting the lower end to a cast-iron flange that could be bolted to a thick, iron cap plate covering the foundation pier and secured by heavy bolts reaching below the street. He said this construction was his own invention. At the top where they were attached to the track the columns branched gracefully into two arms like the limbs of a tree. The effect overhead was not unaesthetic. Whenever necessary to prevent oscillation, he erected a second set of columns at the building line for bracing.

The cable, running in a wooden tunnel-box between the tracks, was carried by small, four-wheeled trucks called travelers, which were attached to the cable every hundred and fifty feet. Each traveler had a horn (or shank) at the top that projected up through a slot in the tunnel-box and made contact with a grip on the car. The method of contact between the cable and the car was what Harvey called his grip-release mechanism, which he had patented in 1866, well before Andrew Hallidie, who is often credited with the invention of cable traction in America. Harvey's was a rather simple plan: a rotating arm controlled by the gripman on the car platform could engage or disengage the horn to start and stop the car. The grip was mounted with a cushioning device intended to soften the shock when the car engaged the cable. Cables were arranged in series every 1,500 feet or so along the line. From the upper track the wire rope would pass through the end column, where it was led into a subterranean tunnel laid under the columns, and enter a vault (where the driving engine was located), pass around the driving wheel and then to the upper tunnel at the other section of the cable. The tunnel-box system worked fairly well except at times during the winter when water seeped into a section of the return tunnel and froze the cable. It is interesting to note that Harvey considered his inventive tasks as fifty percent the adoption of the skeleton iron structure in the streets, twenty-five percent the perfecting of the stationary power attachment to the cable in the vaults, and twenty-five percent to the grip idea.

By June 1868 Harvey had completed the first half-mile section of his elevated, well within the one-year deadline imposed by the governor. During that same month the structure and operation of the railway were carefully inspected by the Rapid Transit Commissioners appointed by Governor Fenton: Freeman Fithian, John Morris, and Jacob Freer. Somewhat later during the month the governor (himself an avid elevated railroad supporter) came down from Albany to take a ride in the new car that had only recently been delivered. He was followed by a host of other nabobs who were curious about the newest thing in urban transportation: Mayor Hoffman came for a ride, then the Croton Board with their head engineer, Craven, followed by the governor of Minnesota, a group from the Common Council of Boston, and several other eminent engineers and civilians. The general opinion of these men was that the el was a great and wondrous machine, a remarkable engineering feat, and a great mechanical success.

Then on July 1, 1868, the Rapid Transit Commissioners made their final report on Harvey's work, giving unanimous approval to his structure and operation. Almost immediately the governor gave the West Side and Yonkers Patent Railway Company his own official blessing, thus vesting in the company full power to proceed with extending the road as far as Spuyten Duyvil at the upper end of the island.

Flushed with pride over their success and eager for their financial rewards, Harvey and his directors took their own test trip from the Battery to Cortland Place two days after the governor's report. Great was their excitment as they found themselves riding at the dizzy speed of twelve miles an hour over their second-story track, propelled mechanically by an engine out of sight and hearing. The cable car was pulled by a three-quarter-inch steel cable manufactured by John Roebling, designer of the Brooklyn Bridge, at his Trenton, New Jersey, works. The cable was made in the same way as ordinary hemp rope, that is with hundreds of fine wires twisted together. The track gauge was four foot ten and half inches, and the rails were on longitudinal girders, no crossties

being used, because they would interfere with the grip connection between car and cable.

No evidence in the newspapers of the day suggested that the trial rides excited much popular interest, and no ceremonies other than the rides themselves marked the event. Today they would be heralded for days in advance and celebrated by some sort of civic festivity accompanied by throngs of reporters. But in 1868 it all seems to have been taken as a matter of course. On Saturday, the day following the trial ride by the company directors, the *Times* did publish under an inconspicuous heading a brief account of the event saying: "The trial trip upon the elevated road in Greenwich street having been postponed on Thursday on account of an accident to the machinery came off yesterday at noon and was very satisfactory. The car ran easily from the Battery to Cortland street, starting at the rate of five miles an hour and increasing to a speed of ten. The chief engineer and inventor expresses the opinion that there is no engineering difficulty in the way of having the railway completed to the Hudson River depot at Thirtieth street the present year. Then the passage from Wall street can be made in fifteen minutes! He is desirous of having the whole line put under contract at once, that the time of its being thrown open to public use may occur during the terms of the present incumbents of the offices of Governor of the State and Mayor of the City, who have recommended the project and assisted its development as a means of relief to the overcrowded thousands of this city."

And so the patent railway appeared a success. Having the utmost confidence in his plan, Harvey had spent most of his savings, some $200,000, to erect the trial section, and he had personally posted the $500,000 bond required for removal of the structure should it not be approved by the governor or state railroad board. Now that the organization of the elevated was put on a firmer basis, he accepted one fifth of the capital stock in the company in return for his services as organizer, use of his patents, and his advance of the $200,000 of his own money.

The inspections and demonstrations having gone well, it only remained for the company to begin regular passenger service and get about the business of earning money. For the next year or two during 1868 and 1869 limited passenger running was attempted, while at the same time the

Carpenters were at work at the first el station at the corner of Day and Greenwich Streets when this picture was made in 1869. It was a modest frame structure squeezed into the second floor of an existing building from which was built a wooden platform cantilevered out over the sidewalk to the track. The ledges provide shelter for the apple vendor and her merchandise. The key sign at the right advertises the services of the local locksmith. COURTESY THE SMITHSONIAN INSTITUTION.

railway worked to extend their road northward up Greenwich Street and Ninth Avenue to Thirtieth Street where they would meet the New York Central Depot, thus offering passengers the attraction of a direct connection from the suburbs in Westchester County to downtown New York.

Travel was restricted to a single track with a number of turnouts located along the main line to permit the passing of cars and to facilitate the transfer of passengers. Cable cars ran singly rather than in trains during cable years. Only two stations were built at first: one at Dey Street and another at Twenty-ninth Street. These first stations were modest frame structures quite unlike the fancy gingerbread Gothic ones of later years. In fact the first station at Dey Street was simply a ledge squeezed into the second floor of an existing building on the corner at Dey Street from which carpenters built a wooden platform cantilevered out over the sidewalk to the track. The station at Twenty-ninth Street was, however, a free-standing wooden shed, quite undistinguished architecturally with a small ticket office. The platform was at track level since the whole station was raised on iron columns identical to those under the railroad. At both stations passengers climbed to the cars on steep, wooden steps. Second-

generation stations on the elevated would be considerably enlarged and more numerous.

The company operated but three rather peculiar-looking cable cars, about thirty feet long, not much bigger than an eight-wheeled horsecar, but low-slung to give passengers a feeling of stability. They had big, open platforms for the gripmen, small windows, side doors, and ogee roofs that looked a good deal like those of a New York omnibus only without a clerestory. Since the elevated was still experimental when the cars were ordered, their design was as plain and cheap as safety would permit. About the only ornamentation on the boxy cars was the heavy European roof cornice and a fresh coat of light-yellow enamel. Though no records remain as to the builder of the cars, it is usually assumed they were the work of John Stephenson of New York City, the biggest omnibus and horsecar builder in the country if not the world. The three original 1868 cable cars did not last long on the elevated. After running for a short time after the changeover to steam locomotives, they were abandoned in early 1872 when new equipment arrived.

Press notices of service on the cable elevated were favorable but hardly what one could consider lavish. *Harpers Weekly* noted with guarded optimism in 1868 that, "The rapid speed attained . . . leads friends of the enterprise to hope that the problem of rapid and safe locomotion through the crowded streets of the city has been solved. It is now in running order and with the present machinery, the cars can be propelled with little jar or oscillation at a rate of fifteen miles per hour." The dignified *Scientific American,* which had given rather little attention to the el during the experimental period, did publish a brief account in their issue of February 1869, saying, "We lately took a ride on the road, and everything seemed to operate well . . . a series of springs at the bottom of the car prevent any sudden shock when the car is put into connection with the cables." The editors added that the centers of the wheels of the cars were made of wood and made little noise. Despite their success in securing the franchise, the life of the Greenwich Street elevated (as it was usually called by New Yorkers) was not without adversity. Cablecars were being operated as far as Thirtieth Street in a regular fashion but in the face of many difficulties: mechanical, legal, and, most of all, financial.

Financing the pioneer elevated had from the start been difficult because most potential backers were naturally skeptical about such an unprecedented mode of travel. A year or so before his death in 1913 Harvey stated that it had been nearly impossible for him to raise money and that he was forced to seek usurious lenders to get the funds needed to keep the project going. The first money for the el was put up by Harvey and his associates in 1867, but once they received official approval from the state it was possible for him to borrow on it. Consequently, in August the property was mortgaged to George Coe and James Benedict for $750,000 to secure a bond issue of like amount. The bonds, issued in denominations of five hundred dollars each, bore seven percent interest to run for fifteen years. The mortgage was signed by W. S. Gurnee, president, and Henry W. Taylor, secretary for the railway. The bond issue was floated by two prominent Wall Street houses of the day: Clark, Dodge & Company and Lockwood & Company. Unfortunately, Harvey, though an extremely able engineer and a skilled inventor, was no businessman, and he never succeeded in getting his elevated further than the experimental stages. When the road fell into financial trouble, Harvey was forced out of his company by unscrupulous men who would later continue the construction and reap enormous financial reward.

Just before Harvey's expulsion from the company, the prospects for the el looked promising; the work of extending the line to Thirtieth Street was in full swing, and investments in the corporation were healthy. Then suddenly "Black Friday" struck on September 24, 1869, creating financial pandemonium. The panic occurred as a result of financial wizard Jay Gould's attempt to corner the gold supply circulating in America. If he could have succeeded in this ploy, he could have set his own price for the precious metal and thus reap titanic profit. Since the nation was on a paper standard, gold was bought and sold as a speculative commodity, the trading taking place in the Gold Exchange at the corner of Broad and Exchange streets in New York. Although the crafty Gould failed to corner the gold market, he had nonetheless made a profit of $11,000,000. In the chaos brought about by his raid, fortunes were lost; Wall Street brokerage houses failed, railway stocks shrank, and the nation's business was paralyzed. One of the banking houses to fail was Lockwood & Company and with it the West Side Elevated's $200,000 fund of working capital on deposit there.

In the emergency to bridge the period until confidence could be restored in the marketplace, Harvey and his directors negotiated a loan of $200,000 to complete the important extension of the railway from a powerful clique of Wall Street stock operators. As a condition for supplying the badly needed capital, the lending syndicate required that the voting power of the majority stock be transferred to them until the loan had been repaid. Later developments clearly indicated that the loan had been made to misuse this voting power. The syndicate then employed an old trick to manipulate the elevated's stock by having the railway appear for a time as a failure while they bought up great lots of stock at low prices; then they would "boom" the shares to high prices by making it look healthy, make large issues of new stock, and then sell—making fortunes in the process. When the syndicate outlined their plan to Harvey, he refused to participate because of the dishonesty to the original stockholders and for its intrinsic shamefulness, even though his share in the profits would have been handsome. Determined to conduct their raid on the elevated stock, however, the clique summarily dismissed Harvey from the company. With him out of the way they felt they could proceed relatively unchecked to carry out their corrupt scheme to exploit the railroad. However, without the guiding hand of Harvey they were beset, in a sort of beautiful poetic justice, with a rash of ills that they could not very easily heal.

One of the reasons for the financial embarrassment of the trestle railroad was the jungle of lawsuits that sought to prevent construction of the remaining part of the track to the New York Central Railroad station at Thirtieth Street. Some of the litigation was brought about by rival street railway companies, who did not fancy having the el as a competitor. There was on Ninth Avenue a surface line, which had been operating on Greenwich Street and Ninth Avenue since 1853. They naturally fought the idea of a competing road passing through their long-established right-of-way, claiming the el was a menace to their rightful patronage. The horsecar companies also inspired many others to legal activities against the elevated. Additional suits were launched by owners

Charles T. Harvey, inventor and builder of the pioneer elevated railway on Greenwich Street. COURTESY THE NEW YORK HISTORICAL SOCIETY.

of real estate along the line who feared the strange, elevated contraption. Property owners had not at first very much objected to the el when it was in its experimental stage, perhaps because they did not take it very seriously. But later when residents came to realize the operation might be permanent, they turned to the courts for redress. Also, whereas the first half mile of the trestle ran primarily through commercial neighborhoods in lower Greenwich Street, the road as it was built northward encountered residential blocks where families living in single dwellings bitterly opposed the intrusion of the railroad outside their doorsteps.

In rebuttal to the property owners one spokesman for the elevated had a particularly sanguine reply. When a group of residents on upper Greenwich Street raised a rumpus that the el cut off so much light from their houses, this man, a lawyer for the railroad by the name of Lawson Fuller, came up with the ingenious but specious argument that New Yorkers should rejoice for having been blessed with a wonderful open-air sanatorium where they could enjoy the benefits of fresh air merely by paying for a ride on the el. Although opposition to the overhead road was not as heated as it would be a few years later when a citywide elevated network was planned using dirty steam locomotives rather than the relatively clean cable as motive power, it was nevertheless a constant hindrance and expense to the infant Greenwich Street el.

More painful and certainly more disastrous than the external wounds of the lawsuits, however, were the company's internal ills, which ate like cancer at their corporate health and morale. Foremost among their troubles was the cable-traction method, which was simply not working satisfactorily. Though potentially an excellent idea, being inherently quieter and cleaner than orthodox steam-locomotive motive power, the cable system had many fundamental defects and weaknesses. After expending large amounts of money in experimenting with it, the new directors of the company came to realize that the method was impracticable. Meanwhile the men who had contributed great sums toward the construction of the line lost confidence in the project. Perhaps if Harvey had not been forced out by the private band after Black Friday, he could have succeeded in perfecting the cable operation. Who knows?

The cable mechanism was liable to get out of order from the rapid wear of the working parts, causing frequent breakdowns that left passengers marooned over the street. The wire ropes that pulled the cars had to operate at such speed and produced so much friction that the grips wore out quickly and the cables frayed. When a cable snapped, there was nothing for the passengers to do but wait to be towed to safety, for there were no platforms except at the stations, and the tracks were balanced to a single row of columns inaccessible even to daring men who might have been willing to shinny down to the street. Passengers also complained of the violent jerks of starting the cars; such head-jarring concussions were not only painful to the rider, but also played havoc with the apparatus as well. Another source of trouble was the complaints from residents along the route about the noise. Far from being silent in its operation, as some early reports had suggested, the running of the cable trucks in the hollow, wooden box under the tracks caused a continual resounding clatter that became to horses and pedestrians alike a source of unending terror. In fact the noise eventually earned for the el the epithet the "rattletrap line."

And then the inevitable happened—an accident. Though not really serious, the misadventure further undermined public confidence in the el. As reported in the *American Artisan* the wreck occurred on June 14, 1870, at the Twenty-ninth Street station of the Greenwich Street Elevated Railway,

The Greenwich Street cable elevated in 1869, where it crossed Little West Twelfth and turned into Ninth Avenue. The structure was heavily braced at this curve. The enclosure overhead housed the running cable that propelled the cars. COURTESY THE NEW YORK HISTORICAL SOCIETY.

"that peculiarly unfortunate example of perverted engineering instincts." As a gripman tried to detach a car from the cable, the mechanism failed, and the car lurched ahead, smashing into another stopped car, driving it about a half a block down the line, and snapping the cable in two. Damage was but slight, though the thirty passengers were frightened and shaken. Since they were stranded away from the platform, they were obliged to scramble to the ground by a ladder. Just the day before, the cable had stopped because of steam-engine failure, causing alarm among the passengers on the line. One old Irish woman was so badly frightened that she had to be carried from the stranded car in a carpet stretcher and taken to the ground via the second-story window of an adjacent house at No. 339 Ninth Avenue. Actually, the cable broke so frequently that a team of horses was kept on standby to pull the powerless car and pasengers to the end of the line. Following such a breakdown the whole system would have to be shut down for a day or so while repairs were made, much to the laughter of the city's engineering community.

With all of the flaws, breakdowns, and interruptions to service, there arose a general condemnation of the elevated by the New York press corps. The *Herald* of May 21, 1870, called for the removal of the structure because "the crude contraption

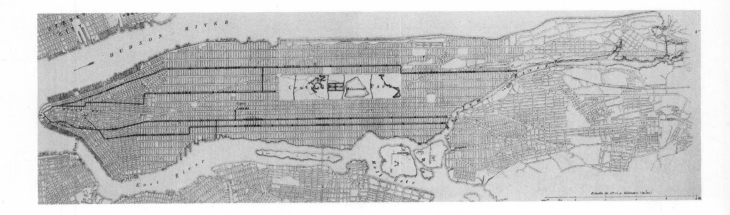

violates all the principles of engineering and all natural laws besides. Let it therefore be done away with at once." Another manifestation of the popular outcry against the stilt railroad was the grand jury called to investigate the many complaints lodged against the safety of the structure. It was becoming obvious that there were grave defects not only in the method of propelling the cars, but also in the strength of the structure as well. Some engineers testified, perhaps with some exaggeration, that the flimsy-looking, one-legged bridge was about to collapse. One witness declared upon his oath "the elevated to be in an unsafe condition, and prejudicial to the lives, comfort, limbs, and personal security of the people passing through the streets . . . and the structure with its 500 iron posts and sleepers and rails an annoyance and a nuisance." Yet being a grand jury the court could not really stop the road nor halt its expansion. After the inquest, however, the company recognized the weakness in the fifteen-inch Phoenix columns and set about strengthening them with braces. Thereafter the superstructure construction was made considerably stronger.

Actually, there was no need for any outside group to stop the elevated. The barrage of opposition was only an ostensible symptom of the disorder within the company. By fall of 1870 cable-car operation was discontinued, and the line remained idle. The el was dead. Abandoned and lifeless, the structure grew rustier day by day, a forlorn and melancholy monument, apparently, to visionary engineering schemes and unwise condition. Dwellers along the line could once more enjoy their privacy undisturbed by the prying eyes of el passengers and the mysterious apparatus of clattering cables. The screeching wheels no longer

terrorized passing horses. Instead of complaining about the elevated railroad, New Yorkers discussed the exotic expedition that the *Times* had financed to have Henry Stanley find David Livingston in British East Africa. In late August the *Times* unnostalgically hailed the passing of the el by saying, "The Elevated Railway in Ninth Avenue has ended its precarious existence. The amusement of running empty cars up and down the road, while it gratified the pride of the inventors and excited the interest of the general public, was apparently too expensive to the stockholders to be longer indulged in." The road died of want of confidence. Its appearance was against it to begin with, and the one or two serious accidents and the many annoying stoppages that occurred during its brief career were not reassuring.

After lying idle for several months, the pioneer elevated railroad, bankrupt and moribund, was sold at a sheriff's auction for $960 to Francis Tows, the property including the railroad from the

West Side Elevated Railway
November 14th. 1870.

TIME TABLE.

Cars leave 29th Street Station 6.30 A. M. and at hourly intervals until 5.30 P. M.,

Arriving at 23d Street Station 2 minutes later.
"	"	Bethune	"	6	"	"
"	"	Houston	"	30	"	"
"	"	Franklin	"	15	"	"
"	"	Dey	"	20	"	"

Cars leave Dey Street at 7 A. M., and at hourly intervals thereafter up to 6 P. M., and Way Stations at above difference in time.

Fare, 10 Cts. Artisan Trains, 5 Cts.

RULES.

To 'accustom' Brakesmen and Engineers to the regular routine operating of the Railway, full stops will be made at the Way Stations, although the platforms are not yet provided for regular use, but will be soon.

One car will leave five minutes in advance of regular time if all seats are occupied, and the last car will leave five minutes after time, if vacant seats then remain. Each train of three cars has accommodation for 120 passengers.

Cars leaving before 7 A. M., and next after 6 P. M., will be known as ARTISAN TRAINS at HALF FARE, or 5 cents, until further notice.

REMARKS.

Over 20,000 passengers have already been conveyed on this Railway *without injury to one of them.* This result has been due to the great care in testing the structure and proving it safe beforehand.

Its safety having been unquestionably proven, its certainty and celerity only remain to be perfected by regular use. Every fare paid encourages improvement in these regards, and all who desire quick transit in New York City are invited to aid in the rapid perfection and certain success of this *pioneer enterprise.*

T. GEREHART,
Acting Superintendent.

Battery to Thirty-first Street, and three passenger cars. (Tows represented the bondholders who were attempting to get the road in running order again.) In the meantime a new company was seeking permission to run steam locomotives instead of cables to pull the cars. Since the original charter had specifically called for cable operation, they applied to the Transit Commission for authorization. Early in 1871, the commission permitted the company to abandon old methods of operation and to use steam engines or dummies for future operations. Consequently, the wire-rope system was removed and all the apparatus, cables, stationary engines, equipment, and machinery scrapped.

4 In Sunlight and in Shadow: Early Steam Years, 1871 to 1877

In the early morning of April 20, 1871, a strange, low rumbling shook residents of the lower West Side out of their slumber. Throwing open their shutters to see what was going on, householders watched as a peculiar locomotive throbbed by their bedroom windows pulling a string of railroad cars. The el was alive again, that morning inaugurating new service with a glossy, black dummy engine, a daintily decorated little machine all neatly boxed in to prevent its scaring horses. Everything seemed wonderfully auspicious for the elevated that spring morning as it made its comeback after the years of failure and near failure. Everything was sanguine—everything that is except the timing. For on the very day of the inaugural public steam-run with the pretty, new locomotive there occurred the bloodiest of New York's perennial Irish riots. Near the elevated's Thirtieth Street station a mob of Irish Catholics attacked a group of Orangemen on Eighth Avenue, who were parading to celebrate the Battle of the Boyne. Murder and mayhem spread through the neighborhood, at one point threatening to incinerate the el itself. By the time the militia had been called in to quell the massacre, the violence had become uncontainable, and the soldiers were forced to fire into the mob. Altogether fifty-two people were killed and many more maimed in the melee.

Despite their ill-timed debut, the new steam locomotives did a great deal to revive the el, and they began running in regular service almost immediately. During the winter the men in control of the elevated had organized a new company, renaming it the New York Elevated Railroad Company. In searching about for new backers they interested a wealthy coal dealer associated with the Pennsylvania Railroad by the name of

The Brooklyn, *one of the first steam locomotives to run on the el, and two shadbelly cars pose along with the train crew in the rapidly developing West Sixties. In those days the expansion of the elevated railroad was quickly followed by commercial prosperity and real-estate development.* COURTESY THE SMITHSONIAN INSTITUTION.

William Scott, whose money and influence it was hoped could put the struggling enterprise back on its feet. Several prominent businessmen were named to the board of directors, men who had ambitious plans for the el. They wanted to add a hundred and sixty miles of trestle to the railway, carrying el service into every ward of New York City: adding new East Side lines, West Side lines, crosstown spurs, and suburban runs to Westchester, Yonkers, New Rochelle, and Tarrytown. Stock in the new corporation was issued to the amount of $801,825, but Charles Harvey's name was not mentioned anywhere in the reorganization.

Aside from a few snags, such as whether the newly organized elevated could legally inherit the right to use steam locomotives from an earlier company, the road seemed to be healthy and prospering. Having replaced the unsatisfactory

cable with steam locomotives, strengthened their structure, attracted new capital, and improved their public relations, the elevated looked forward to a happy and profitable future. But after overcoming so many formidable obstacles, they were threatened with yet another antagonist in the form of William Marcy Tweed, one of the most infamous corruptionists in the history of New York City and political dictator of the city in the post-Civil War decade. Boss Tweed had a plan for a rapid-transit system of his own, and he set about to discredit and destroy the New York Elevated, his principal rival in the field. With the vast resources at his disposal from graft and corruption he very nearly achieved his goal.

Tweed's power in 1871 was complete, and his influence was at its zenith. He held many powerful state and city offices: School Commissioner, Assistant Street Commissioner, President of the Board of Supervisors for the County of New York, Senator of the State of New York, Chairman of the Democratic Central Committee of New York County, and Grand Sachem of Tammany Hall, the title that pleased him most. And as Superintendent of Public Works for the city, he authorized and

45

An early morning el train runs close by the still-shuttered windows of a row of pretty Greek revival houses in New York's Chelsea district. To the right is the road's Eleventh Street station and a turnout for trains. This photograph was taken by A. C. Johnson, conductor of the train, in 1872.

planned all new public-works programs, a source of unimaginable graft. In fact, under the city's new charter of 1870 Tweed actually had more power than the mayor. He was also one of New York's principal landowners, was on speaking terms with the rich and powerful, and did business with Jay Gould, Jim Fisk, and Cornelius Vanderbilt. Judges rendered decisions according to his requests (he is reported to have owned Judge Albert Cardoza). Legislators in Albany passed or defeated laws as he determined; New York Mayor A. Oakey Hall did just about exactly what the Boss asked, and the same went for Governor John Hoffman, who was Tweed's stooge. Tweed was not sparing with his gifts, especially to those rich and powerful men who were flocking to New York to make stupendous fortunes in that wild and reckless postwar era. Jay Gould, who would later own New York's

elevated railroads, engaged Tweed's aid in 1868 during the great battle with Cornelius Vanderbilt for control of the Erie Railroad. Boss Tweed used his influence with New York judges to secure the passage of a law altering the method of electing Erie's board of directors in Gould's favor. Tweed's Judge Cardoza also protected Gould in 1869 from paying his debts after his attempt to corner the gold market, the coup that led to Black Friday and coincidentally Charles Harvey's loss of his elevated railroad. Later, in return for getting special privileges for the Erie to transport westward-bound immigrants from the reception center at Castle Garden, Gould appointed Tweed to the Erie board.

With such titanic financial and political leverage at his disposal Tweed was able to control the press to his satisfaction. Several reporters on each New York daily received stipends to ensure favorable coverage, all smiling on the Boss and giving readers the impression that everybody admired the city government. Though history's judgment of Tweed is almost universally scathing, there is no doubt about the man's shrewd political acumen. He had a genius for organizing things and he made politics pay. He had the reputation for being intensely loyal to his friends and could rally diverse groups in polyglot New York City to his programs, encouraging labor-union support and Roman Catholic allegiance, though he was himself a Scotch Protestant. Measuring the dimension of his thievery taxes the imagination. Precisely how much he and his cohorts managed to steal from the people of the city will never be known for certain, but estimates range from seventy-five million to two hundred million. His pattern for graft-taking was extremely efficient. He and his henchmen allowed contractors for schools, streets, parks, and other public works of all sorts to overcharge the city, and then they received a kick-back on the lavish profits. Between 1869 and 1870, for instance, the city spent a total of $3,200,000 on armory repairs alone, though later the *New York Times* estimated that twelve out of every thirteen dollars were pure graft for the Tweedring.

With his keen sense in exploiting every opportunity, high or low, big or little, that suggested a chance for plunder it is no wonder that he seized upon a scheme for rapid transit, because it offered such a bountiful harvest in graft. Well aware of the need for a solution to the traffic problems plaguing the city, he advocated the so-called Viaduct Plan, a sort of grandiose elevated railroad running over a masonry viaduct through the length of Manhattan, supported by huge, stone arches forty feet high. The New York Railway Company, as it was officially called, was incorporated by Tweed and others in April 5, 1871. Graft involved in the construction of the railway alone would have been enormous, probably millions per mile, but the idea was made even more lucrative by the provision that all the property in the path of the railway be condemned and razed and that the streets along the course of the road be widened and graded. The viaduct scheme was to be given the right to build other lines in any part of the city, thus virtually enabling the ring to put a railroad in every street of the city if they chose. To get the project going, the city would have been compelled to buy five million dollars of stock; this extortion would have been just a start, for Tweed would have followed up by asking more money from the Legislature when the five million had been divided up among the Tammany ring.

To disguise the true intent of his scheme and to give it an air of respectability, Tweed asked a number of wealthy New York capitalists to act as incorporators of the viaduct railroad. They seemed an unimpeachable band of distinguished men: John Jacob Astor, Horace Greeley, Oswald Ottendorfer, the saintly Peter Cooper, August Belmont, Charles A. Lamont, Levi Morton, Charles Tiffany, John Agnew, Simeon Chittenden, and James Gordon Bennett, Jr. A magnificent suite of marble offices was rented at Broadway and Warren Street, and printed plans were published showing a grand terminal facing on City Hall Park.

The only rival in sight was the struggling New York Elevated, and this Senator Tweed with typical cunning decided to discredit and destroy. Having already stopped the novel subway that Alfred Beach had recently built near City Hall, Tweed next pushed through the Legislature a bill branding the el as a public nuisance and permitting him as Commissioner of Public Works to tear it down within ninety days. According to the New York *Herald* the corruptionists boasted that they would not only tear down the road, but would fine and imprison the backers as well. Engineers and newspaper editors were hired to assert that the Greenwich Street el would not stand, was dangerous to the lives of passengers, would cause constant runaways of horses, and would destroy business. Some attempts were even made at one time to incite mob violence against it. The directors of the New York Elevated were taken by surprise by the Tweed attack in Albany, but fortunately Charles Harvey was informed of the emergency. Although he was totally out of the company at the time, his immediate response was in the public interest and in saving the elevated rather than personal revenge upon the men who had ruined him. As soon as he could, Harvey got in touch with his friend and powerful ally in Albany, Erastus Corning, whose aid he solicited to save the elevated for friendship's sake, since Corning had no investment in the road. Corning had considerable influence with the legislators from the upstate counties and was considered as the most influential citizen of the state outside of New York City. When the vote was taken on Tweed's bill to tear down the el, it was defeated by the decisive majority of seventy-four nays to thirty-four yeas, the only defeat dictator Tweed experienced during that session. But for Corning and Harvey's interference in defense of the el, the railway would

Through Battery Pak passengers on the elevated rode in the treetops and could look down on picnicking families. Battery Park station has just been completed; in fact, a workman's ladder is still propped against the trestle while a dummy engine and two cars rumble along en route to South Ferry.

have been destroyed and the system too discredited to enlist new capital for reconstruction. With Tweed in control the viaduct system would doubtless have been adopted as the only transit system available. However, as it turned out, Tweed's defeat in Albany spelled the downfall of the viaduct plan, and nothing more was heard about it, thus sparing New York one of the grandest swindles of all time.

Actually, it is a wonder that the viaduct plan was not successful, considering the Tammany leader's power and influence, but apparently it was too late in coming and got caught in the collapse of the Tweed empire. Tweed's failure with the bill was an omen; his days were numbered. For more than a year caricaturist Thomas Nast had conducted a brave and brilliant attack on the Tweed ring in the pages of *Harpers Weekly,* exposing their wickedness and implanting the image of Tweed as a cartoon figure in the public mind for generations to come—a gross, half comic character of corruption, an obese, lecherous Falstaff. Reform was in the air, and in December 1871, just months after this attack on the elevated, Tweed's past caught up with him. Indicted and arrested as a common felon, he was to die ignominiously in Ludlow jail in 1878, a despised and broken man. After the shock of the Tweed scandals, New York became more conservative and a good deal more honest in its government. The change was to have an effect on the elevated railroads, for in the future construction would have to be financed entirely by private moneys. There would be no more thought of rich public subsidies as in the old days, and the notion of the subway was discredited because it was thought far too expensive. A reaction had set in in the city, and the government was much more economy-minded than during Tweed's big-spending days. The elevated lines constructed in the second half of the 1870s were done by vigorous entrepreneurs using the instruments of the government only for planning and inspection, not for public capital. The only public assistance the elevated nabobs received was use of the streets, and some critics thought even that unfair.

Tweed and the viaduct plan vanquished, the elevated prospered and grew prodigiously. By 1872 passenger traffic had become brisk enough to cause the company to buy additional locomotives and cars, build several new stations, and begin to push their tracks farther northward. Riding on the el was by this time growing rather popular with New Yorkers, many of whom had gotten over their early fears of such aerial travel.

The elevated in 1872 was running twenty-eight round-trip trains daily between Dey and Twenty-ninth streets with three or four cars, the three-mile trip taking about fifteen minutes, which was extraordinarily good travel time considering that the wretchedly crowded streetcars took more than twice that. Feeling rather secure about their operations, the directors published a timetable and attempted to coordinate their schedules so that el trains would connect with those of the New York Central Railroad, which left from their Thirtieth Street depot throughout the day for Yonkers and the numerous villages along the Hudson River south of Sing Sing. The *Railroad Gazette* that August observed that morning and evening trains of the New York Elevated were pretty well filled with commuters who used the trains every day. A good number of them were from Yonkers, changing from the New York Central trains to the el to travel to their offices in Wall Street. According to the timetable a passenger leaving Yonkers at 7:15 in the morning arrived at the el's Dey Street station at 8:10.

The expansion and regular service of the el marked the first real advance in the city's rapid-transit system, and travel on the elevated began to take on a certain degree of prestige. There can be little doubt that it was the quickest and pleasantest way of getting around. The cars traveled high enough above the streets to give passengers a good perspective of the city and the dancing waters of the Hudson River to the west, and they were out of the nasty dust and smarm of the pavement. Unlike street travel, the elevated track had no obstructions from trucks and wagons, which were so annoying in the congested downtown roadways. One English visitor to New York said he liked the elevated better than his own London subway because of the remarkable view, and a German geographer who rode on it in 1873 said that with the el New York's transportation facilities were better than those of any European city. Little or no complaint was heard from people living along the line. According to the *Railroad Gazette* the little locomotives burned anthracite and made little smoke, nor were they very noisy, though the escaping steam was said to be noticeable in the street under the tracks. Trains shuttled constantly over the trestle, usually over streets crowded with horse-drawn vehicles, though most of the animals were so accustomed to city noises that they took little notice.

As traffic increased, the company built new stations along their route, for the original two depots were inadequate. In 1872 new stations were opened at Watts Street, Little West 12th Street, Morris Street, and at Twenty-first Street and Ninth Avenue. In the following year new stations were built at Franklin Street, Houston Street, and in a building known as No. 7 Broadway opposite the Bowling Green with the railroad station on the ground floor and the offices of the company above. The two original terminals at Dey Street and Twenty-ninth Street were shut down and new larger ones built at Thirtieth Street and at Cortland Street. While these second-generation elevated stations were modest frame structures without waiting rooms, they nevertheless offered passengers better and closer service than before.

As the company prospered, the directors felt flush enough to buy some new rolling stock: an additional dummy engine and four cars. The new passenger cars were of a rather unusual design, the bodies being depressed between the tracks for stability so that the center of gravity was brought as close as possible to the rails. This "shadbelly" design, as it was called, was adopted for safety purposes to dispel any worries of passengers who feared that the train might fall off the trestle. Since the floor was only four-and-a-half inches above the rails, the cars were very stable; it would have required a substantial shock to overturn one of them. The low-slung design also created some interesting features inside. The depressed centers of the cars meant that the ceilings were quite lofty (nine feet six inches), creating a remarkable spaciousness. They were also unusually well lighted for the time because the builders, Jackson & Sharp of Wilmington, Delaware, had provided a double course of windows in the dropped center portion, no doubt providing passengers an excellent view of the city in transit. The cars were thirty-five feet

49

In 1876 the New York Elevated extended its double track service up Ninth Avenue as far as Sixty-first Street. This photograph shows the official opening of the new track at the Forty-second Street station. A crossing bridge connected the two tracks and the waiting room, a Victorian bungalow with crisp, ginger-bread trim. Looking on are many darkly dressed men, no doubt investors or directors in the company. COUR-TESY THE COLLECTION OF GEORGE RAHILLY.

long, weighed 9,400 pounds, and had seats for forty passengers along the sides like those in a streetcar. Doors were at each end with open platforms. They were painted a light-yellow color called "straw" similar to the cable cars. The interiors, we are told by observers, were finished comfortably and handsomely. The only objection seems to have concerned the ventilation, for while the cars had ample provision for exhausting the foul air, there was no way for admitting pure air from the outside.

Enormously proud of their elegant new trains, the New York Elevated Company decided to conduct a bit of public relations to advertise their improved service. Here is part of a flyer that was sent around in the fall of 1872: "We now take and receive passengers at Morris, Dey, Canal, Little West 12th, and Twenty-Ninth Streets. We run four unique elegantly finished and furnished cars, made expressly for our road, capable of seating 44 passengers each, and we take no more than can be seated. We are frequently compelled to refuse passengers after our cars are full. We carry about 1300 passengers daily . . . . We believe we are developing what will enhance that value of real estate, solve the problem of quick transit, relieve our overcrowded streets and sidewalks, be of great public service, and a successful paying enterprise." This was not empty boasting; the company was probably quite accurate in their

50

statement to the public, for improved transit on the West Side of the city was creating a minor real-estate boom on that side of the island. People on the inside were buying up tracts of land early, sitting on it until the elevated could be extended farther up the city, and then selling it off at a fantastic price. The strip that became Riverside Drive should have cost only $1,400,000, but the speculation had driven the cost up to $6,000,000 by the time of its purchase in 1872.

An early "shadbelly" on a flatcar at the builder's plant in Wilmington, Delaware, awaits shipment to New York. These el cars were built with a low center of gravity to overcome passengers' fears that they might tip off the trestle. The dropped centers made room for a double course of windows, which might have given riders a superb view of the city en route–except that in typical mid-Victorian fashion the top row of glass was covered with elegant swagged and tasseled draperies to screen the sunlight. COURTESY DIVISION OF HISTORICAL AND CULTURAL AFFAIRS, STATE OF DELAWARE.

In the fall the Beecher scandal broke wide open, and New Yorkers could think of scarcely little else. The Rev. Mr. Henry Ward Beecher was one of the best-known and best-respected men in the entire metropolitan area, rector of the prestigious Plymouth Church in Brooklyn, and brother of the authoress Harriet Beecher Stowe (*Uncle Tom's Cabin*). No one could have been more shocked when Victoria Woodhull, a lady stockbroker, feminist, spiritualist, and all-around adventuress, branded Beecher an adulterer. It was also rumored that he had been carrying on an affair with one of his young parishioners, Elizabeth Tilton, wife of a prominent liberal editor and poet. By the time the press had tired of the sensation story, the winter of 1872–73 had broken, and the elevated was at work erecting another section of trestle at the upper end of the line from Thirtieth to Thirty-fourth Street. This construction was made under the direction of the company's very able superintendent, D. W. Wyman, who had also helped to

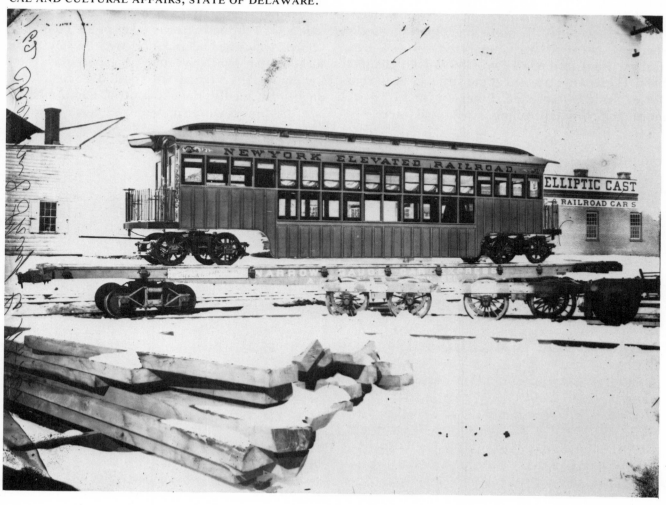

design the first el steam locomotives and the "shadbelly" cars. Many authorities believe that after the dismissal of Charles Harvey it was Wyman's engineering skill that brought order out of chaos. From a railroad that was almost universally pronounced a failure, he had developed a line that steadily grew in public favor and demonstrated the practicability of the elevated railroad system to the entire world.

Since the old portion of the elevated structure inherited from the first generation days of cable power was thought to lack stability for the heavier, locomotive-drawn trains, Wyman strengthened the new portion of the line built under his supervision. In second-generation el construction columns were formed out of four round, solid, wrought-iron bars four-and-a-half inches in diameter. At the top the bars were bent outward into four branches, two of which were for lateral support and two for longitudinal stiffness. The four bars were tied together with wrought-iron bands at the neck, or just below the point from which the bars began to branch outward. These columns were placed from thirty to sixty feet apart, and the roadway between was supported on two iron girders, each formed of two twelve-inch channel bars. These were trussed at the street crossing but not in the shorter spans. Later on, in 1875 with more and heavier rolling stock and increased passenger loads, the track gauge was changed from four feet ten inches to standard four feet eight-and-one-half inches. At that time the entire line was closed down for a short period while two hundred workmen changed the wheel widths on the rolling stock and relaid the rails on crossties.

Though under the direction of Wyman the elevated structure was more stable than before, the road was still one-legged (resting on one row of columns), and some people fretted about the inherent danger of riding the overhead trains, fearing that any minor mishap would cause the trains to topple over into the street. As it turned out, however, passengers on the el were quite safe. The principal danger was not with the trains but with the traffic in the streets unmindful of the columns. On one occasion a teamster driving under the elevated's station platform on Morris Street ran against one of the supporting columns, which gave way, bringing the platform down and crushing him to death. Another accident occurred a few months later. John Strickland had just bid his wife goodbye to start off on a season's tour in charge of Barnum's circus team of six coal-black horses, which he had under reign at the time. As he turned into Ninth Avenue, a passing elevated locomotive frightened his spirited horses. They dashed off madly, he lost control, and a moment later his brains were sprayed over the sidewalk.

5 Dr. Rufus Gilbert and the Completion of New York's El Empire

Not long after the New York Elevated Railroad began to pull their trains with the sprightly dummy engines, a rival appeared on the scene—the Gilbert Elevated Railway, chartered by the state in 1872, "To operate over the streets, avenues, and thoroughfares in New York City," in tubular, iron roadways straddling the street and supported by gigantic Gothic arches springing from curb to curb. As described in chapter 2, passengers would ride in two great metal tubes through which they would shoot around the city in bulletlike cars by atmospheric pressure. It was a wildly wonderful plan but, of course, totally impracticable. Nonetheless, the government was so eager to complete some sort of transit system for New York that they granted Gilbert a franchise anyway, allowing him a reasonable time to complete his elevated as far north as Forty-second Street. A committee appointed to decide the path the tube-

road should take through the city chose a loop that eventually created the Sixth and Second Avenue elevated routes.

The originator of the novel "Improved Elevated Railway" was Dr. Rufus Henry Gilbert, a physician turned inventor, who succeeded in making a name for himself in the realm of rapid transportation after the Civil War. Born in Chenango County, New York in 1832, young Gilbert began his career as a drug clerk, but his taste for machinery caused him to study mechanics, thus laying the foundation for the knowledge that became so useful to him later on. Having completed his apprenticeship as a machinist, Gilbert for some reason left the trade and went to Corning, New York, to study medicine. From there he went on to New York City to enter the College of Physicians and Surgeons, where his proficiency quickly attracted the attention of the dean, Dr.

Willard Parker, who made him an assistant. After graduating he settled once again in Corning, where he became a popular doctor, specializing in surgery and developing a large and lucrative practice. With the performance of a number of difficult operations his fame spread, though the demands for his services so affected his health that after a few years he was compelled to give up his practice entirely.

Following the death of his wife, Gilbert set off on an extended European tour, occupying his time between cathedrals and castles by visiting hospitals. During the course of these visits, he became deeply impressed with the great number of hospital cases of people living in densely populated tenement districts, particularly in Paris and London. Attributing the cause to lack of sunlight and pure air, Gilbert began to ponder the notion of a cheap rapid-transit system that might take the urban poor out of their dark rookeries into the salubrious air of the country and thus serve as a means of improving public health. After partially regaining his health and eager to work out the engineering details of his idea, the doctor returned to New York, where the Civil War had just broken out. Feeling that his skills as a surgeon would be badly needed, he enlisted in a regiment of Zouaves. After serving in the field for several campaigns, he was made medical inspector of Fortress Monroe, then medical director of the XIV Army Corps, and later medical director and superintendent of the United States Army Hospitals, a post that he held until the close of the war, when he resigned and returned to his home. Rather than taking up his medical practice, Dr. Gilbert accepted a job as assistant to Josiah Stearns, then superintendent of the Central Railroad of New Jersey, where he had a good opportunity to study the problems of transportation at close range. As soon as his work for the railroad was completed, he resigned to devote his whole time to his rapid-transit scheme. The first hurdles in his new career were crossed rather smoothly; he obtained patents for his pneumatic tube system in 1870 and received a franchise to build in 1872. His greatest obstacle was in raising the capital to construct his elevated. The financial depression following the panic of 1873 together with the impossibly expensive method of construction and operation he had outlined prevented his company from getting off the ground.

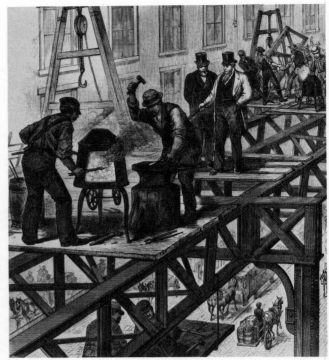

Workmen of the construction force of the Gilbert Elevated Railroad are heating rivets for bracing the ironwork on the Church Street section. Financiers Russell Sage and Cyrus Field inspect the progress of the construction in the spring of 1878. COURTESY Frank Leslie's Illustrated, APRIL 27, 1878.

For four years Gilbert sought financing for his weird tubular elevated with but little success. Then in 1875 the state passed an important piece of legislation, which not only created a healthy investment climate for Gilbert's road, but also cleared the way for a comprehensive citywide elevated-railroad network. The Husted Act empowered the mayor of New York to appoint the city's first Rapid Transit Commission to determine the most appropriate type of transit system (elevated, subway, or whatever), locate the most advantageous routes through the city, and determine the best plans for construction. The act was particularly notable in that for the first time the future of New York's public transportation was put in the hands of a city constituency rather than the Albany-based politicos. The elevated railroad system in metropolitan New York that was eventually built owed its existence to this city Board of Commissioners of Rapid Transit, who provided the building of the Gilbert road and laid down the expanded and connecting routes for the New York Elevated, resulting in the lines on Second, Third, Sixth, and Ninth avenues.

In general the Husted Act provided that, whenever fifty or more property owners in the city made a formal application for the need for a railway in their neighborhood, the mayor would appoint a body of five commissioners to study the request and, upon approval, fix the routes over which the road would be built; they were also to make any connections between the new road and existing elevated lines, ferries, and railroads that were thought necessary to expedite travel. In mid-1875 two petitions from householders were presented to Mayor Wickham, one asking for another steam railroad in the city and the other requesting that the New York Elevated be extended to connect with other railroads and ferries. Acting on his authority the mayor appointed five men to decide on these requests and to report their findings in early 1876: Joseph Seiligman, Lewis B. Brown, Cornelius Delamater, Jordan L. Mott, and Charles Canda.

Deliberating over the winter, the committee made their recommendations on the future of New York transit in the spring. The year was a very special one, the country's one hundredth anniversary, and the Empire City was bursting with a remarkable energy as though to prove to the rest of the nation its special vitality and progress in the centennial year. On the East Side of the city the New York Elevated continued its steady march of new track toward Fifty-ninth Street, uptown, close by the just completed Central Park. On the West Side the mammoth Gothic towers of John and Washington Roebling's great Brooklyn Bridge were soaring high over the East River, nearing completion. In the center of Manhattan in Madison Square the uplifted hand of Bartholdi's statue of "Liberty," gift of the people of France, was put up on display. To add further to the excitement and hurly-burly, Professor Huxley visited the city in the summer, shocking the Victorian sensibilities of Protestants and Roman Catholics alike with his lecture on the theory of evolution. New Yorkers who could afford it took the train to Philadelphia to visit the splendid Centennial Exhibition, which had been opened by President Grant himself. There on display, among so many other triumphs of American industrial progress, was one of the New York Elevated's plucky little dummy locomotives, appropriately lettered *Centennial*. To cap the remarkable year of 1876 was a presidential election, held of course in November, and

casting two contenders for the high office: Democrat Samuel J. Tilden of New York City and Republican Rutherford B. Hayes.

The Rapid Transit Commission's first decision was to choose the elevated railroad as the most advantageous system for moving New Yorkers around the city speedily and safely. There were next faced with the task of deciding the routes the two existing el franchises should take. The work of fixing routes was complicated by a number of city ordinances that prohibited public transit in certain streets, especially the one that banned the elevated from even crossing Broadway south of Thirty-fourth Street. The commission eventually got around this legal difficulty by extending the New York Elevated's structure south through Battery Park to South Ferry, then looping it northward and running up the east side of the island along Third Avenue to Forty-second Street to connect with the Grand Central Station of the New York Central Railroad and beyond to the northern reaches of the city. At the same time the commission approved the routes for the Gilbert Elevated on Sixth and Second Avenues.

They also advised and practically forced Gilbert to adopt a simpler and more economical plan for construction than the fanciful and horribly expensive Gothic arch structure and to change the mode of operation from the dubious and unproved atmospheric air-pressure system. The body was clear to point out that the practical use and experience demonstrated by the steam locomotive, now a well-tried machine, was better adapted to elevated transit than the unproved system of compressed air. Since Gilbert had failed to raise funds to build his el, this ruling by the commissioners was no doubt a blessing. Though the Gilbert Elevated as it was built had very little resemblance to the inventor's original patent, he did retain the valuable franchise and routes through the city. The concession to Gilbert allowing him to change his construction plans and method of operation, along with the priceless routing through the center of the city, placed his company in an enviable position. Consequently, after his scheme had lain dormant for so long, promoters and backers rose up immediately to finance his railroad.

Like Charles Harvey, Gilbert lacked the financial prowess to promote his elevated with success. In his efforts to enlist capital, he gave away control

Chatham Square, where the Second and Third Avenue lines and the City Hall spur converged, was always a busy junction on the elevated. This photograph shows the new Second Avenue el station when it was just completed in 1880. Behind it is the overhead footbridge, which allowed passengers to change to the Third Avenue station for transfer to trains for City Hall and the Brooklyn Bridge. COURTESY THE MUSEUM OF THE CITY OF NEW YORK.

of his company and was eventually frozen out. Gilbert had unquestionable vision but was clearly not an entrepreneur. Realizing the profits inherent in Gilbert's valuable and densely populated route, swindlers watched like vultures, waiting until he was desperate before they struck. The population of the Sixth Avenue route that ran up through the middle of Manhattan Island was greater than the West Side route of the New York Elevated, whose territory was not only smaller, but had more area devoted to large warehouses and stockyards. In his attempt to secure financial backing, Gilbert fell in with an unscrupulous and conniving man, William Foster, who promised the support of the New York Loan and Improvement Company to finance and construct the line. Distraught and approaching desperation after four frustrating years, Dr. Gilbert was willing to listen to anyone, and he soon agreed to reorganize the company, creating new directors, taking Foster as a partner, and issuing $3,500,000 in stock. Foster took up thousands of shares to sell, but he actually did not sell any at all; instead he formed a ring with J. T. Navarro and the Loan and Improvement Company, who soon acquired entire control of the company, including the stock Gilbert had counted as his own. Dr. Gilbert, occupied with the hundreds of technical problems of his construction work and joyous with the idea of seeing his elevated actually going up, never suspected what was happening to him.

Going back to the work of the 1876 Rapid Transit Commission, the men had set fares on both el lines from the Battery to Fifty-ninth Street at ten cents; from the Battery to Harlem no more than fifteen cents, and to Highbridge not more than seventeen. Their resolutions also provided for special commission trains to be operated for the working people of New York with fares half the regular price from 5:30 to 7:30 in the morning and from 5 to 7 in the evening. As a result, what are usually called rush hours in most cities were known for a long time in New York as commission hours because of the special fares authorized by the Transit Commission.

Amidst the involved and tedious work of getting the elevated system operating in some coordinated and systematic way, the commission had also established a separate railroad company, which, while being a further complication among so many other details of the corporate history of this period, is important to the future of the el. Since the elevated system was believed to be so important to the city, they felt that in case neither the New York Elevated nor the Gilbert Elevated succeeded in completing their roads as scheduled, there ought to be some sort of backup company who could then step in and finish the construction work and operate the road, subject to the same rules and regulations as the other companies. The Manhattan Railway Company was created for just this purpose. However, the commission declared specifically that it should not be brought in unless one of the two elevateds failed to meet its obligations. The incorporators of Manhattan included men who were very much interested in getting control of the New York Elevated or the Gilbert Elevated should they run into difficulties: Cornelius Garrison, Horace Porter, Milton Courtright, John Tracy, George Pullman, Jose Navarro, William Scott, David Dows, and John Ross.

With all of the attention and special consideration given by the Rapid Transit Commission to the Gilbert Elevated, the long-established New York El began to feel somewhat slighted. Since they were the pioneer in elevated transit and had waged such a long and difficult struggle for survival, expending great sums of money in building and equipping their road to demonstrate its success, they felt they were entitled to more favors from the authorities. In 1875 that company was running cars in regular service up Ninth Avenue and Greenwich Street as far as Thirty-fourth Street and planned an immediate extension to Fifty-ninth Street. Receipts from passenger fares for the year were $82,945, a remarkable figure considering that the trains did not run on Sundays. Under the circumstances the New York Elevated believed

that the Gilbert line was being given unfair encouragement; however, as it turned out, the New York El, because of their experience in operations, were well ahead of Gilbert and enjoyed a very favorable competitive position. About this time Cyrus Field, of Atlantic cable, invested heavily in the New York Elevated, using much of his fortune to purchase control of the system. His reputation and moneys gave a great stimulus to the whole undertaking and did much to assure the success of the company and promote its expansion.

The work of the Rapid Transit Commission of 1875–76 had an enormous effect on the city, for almost immediately the two roads proceeded to construct their trestles under the committee's authority. Rail by rail and pillar by pillar the el began to take shape on broad avenues and in narrow streets where the aerial line was built right over the sidewalk and almost flush with the sides of buildings. The work of the commission assured a certain and fast solution to the transit question, and by fixing only two basic lines created for both the New York El and the Gilbert line the likelihood of great success. In the fall of 1876, while both companies were pushing ahead with their construction, lawsuits were initiated from property owners and horsecar companies, who declared the Husted Bill unconstitutional and the action of the Rapid Trust Commissioners illegal. These injunctions completely halted all construction work for nearly a year while the courts debated the difficult issue. Finally, in November 1877 the Court of Appeals ruled in favor of the elevated, thus removing all serious legal obstacles that had been impeding new construction uptown and on the East Side. The injunctions having been dissolved, the New York Elevated and the Gilbert Elevated roads moved swiftly through the remaining years of the decade to complete the badly needed citywide system that ran on four great iron bridges lengthwise along Manhattan Island. Following is a brief chronology of the major extensions of the West Side lines on Ninth and Sixth avenues and the East Side lines on Third and Second.

After being granted a right-of-way through Battery Park by the city, the New York Elevated opened its double-track extension in April 1877 with three stops: Battery Place, Battery Park, and South Ferry, site of the splendid, turreted ferry houses serving Staten Island, Brooklyn, and Jer-

The opening of the Gilbert Elevated on June 5, 1878, drew excitement from the people of New York as the first train steamed down Sixth Avenue in from the imposing new Jefferson Market and Police Court. COURTESY PRINTS AND PHOTOS DIVISION, LIBRARY OF CONGRESS.

sey. The double-tracking continued north up both side of Greenwich Street and Ninth Avenue to Fifty-third Street, opening to public use in June. At this time the old structures of former years had been taken down and entirely rebuilt.

Just three days later Gilbert's Sixth Avenue elevated was opened to the public with great panoply. The inaugural was made especially festive since the company invited New Yorkers to ride free on the first day from South Ferry to Central Park; later the fare would be ten cents. The press wrote glowingly about the new service, commenting particularly on the stylishly furnished Eastlake waiting rooms, the decorative ironwork on the staircases, and the covered, scenic pavilions where passengers could enjoy a pleasant promenade while waiting for their trains. Snappy new apple-green cars custom-built for the elevated by the Pullman Palace Car Works of Detroit,

Michigan, were handsomely fitted out with rich, mahogany paneling, Axminster carpets, and comfortable, leather seats. Stations first opened were at Rector, Cortland, Park Place (all on Church Street), Chambers, Franklin, Grand, Bleeker (all on West Broadway); Eighth, Fourteenth, Twenty-third, Twenty-eighth, Thirty-third, Forty-second, Fiftieth, and Fifty-eighth (all on Sixth Avenue). Shortly after the grand opening of the new elevated the directors changed its name from the Gilbert Elevated to the Metropolitan Elevated Railway Company, thus removing all trace of the man who first conceived and incorporated it.

The opening of the Sixth Avenue el on June 5 should have been the happiest day of Gilbert's life; however, on the next day he was voted out of the directorate, then in a few days excluded from the company altogether. Foster and Navarro had taken it all. For seven years Gilbert struggled in the courts, exhausting all that was left of his money in lawsuit after lawsuit in a pathetic attempt to regain some interest in his railroad. But in spite of the extensive litigation he, like Harvey, never received satisfaction. He died at the relatively

Dr. Rufus Gilbert.

young age of fifty-three, a poor and disillusioned man, while the elevated railroad he had done much to create was whirling everywhere about the city achieving a brilliant success. It seems ironic that for all of Gilbert's well-intended dreams that his scheme might improve public health, slum dwellers never really used it to go to the country anyway. In fact the el probably merely created new slums where they had not existed before, make dingy, sunless streets of what had been broad, bright avenues. The failure of the inventor's dream was overwhelming.

During the next year the two West Side els further extended their trestles from Fifty-third to 104th Street and Columbus Avenue (Ninth Avenue), opening eight additional stations about every six blocks. Since the section north of Fifty-third was a common route for both companies, the new double-track structure had been built jointly by the New York and Metropolitan companies.

On September 1, 1879, the Manhattan Railway, which had been created by the commission in 1875, leased both the New York and Metropolitan elevateds, thus unifying all el operations under a single management headed by finincier Jay Gould.

The new operating company hoped the merger would improve the efficiency of the lines by cutting duplicate costs and reducing salaries of some top officials. Moreover, since the two separate companies had to share the same tracks in Pearl Street and in Ninth Avenue from Fifty-third to 110th streets, the merger would ease the problems of scheduling and dispatching. A more reasonable explanation for the Manhattan takeover, however, would seem to be the desire by the syndicate controlling Manhattan Railway—most of whom were known to be shrewd stock manipulators—to exercise power over the New York elevated railroad system and enjoy a profitable monopoly.

At the time of the Manhattan takeover the system consisted of eighty-one miles of track, the New York El having contributed thirty-seven and the Metropolitan forty-one miles. Before the year was out, the Sixth Avenue and Ninth Avenue lines were extended north on Columbus Avenue as far as 110th Street, turning east one block to Eighth Avenue, then north again to 125th Street, with a new station at 116th Street. This structure was the highest part of the entire system, the 110th Street bend often being referred to as "Suicide Curve" because of the large number of people said to have leaped to their deaths there. By December four new stations were opened between 130th and 155th streets, thus completing the West Side elevated to the uppermost reaches of the island at the Harlem River. The terminus of the West Side lines at 155th Street was shared a few years later by the New York and Northern Railroad, which came across the river from the Bronx. The terminal station became a convenient transfer point for passengers to change from points in Westchester County to the elevated service on Manhattan. Just north of the terminal Manhattan Elevated built their extensive yards, including a 1,500-foot waterfront for coaling operations. The early home of the New York Giants baseball team, the Polo Grounds, was also built close by.

The New York Elevated's East Side line on Third Avenue opened in August 1878 from South Ferry to the Grand Central Depot, via State Street, Front Street, Coenties Slip, Pearl Street (with stations at Hanover Square, Fulton Street, and Franklin Square); New Bowery, the Bowery (with stations at Chatham Square, Canal, Grand, and Houston). On Third Avenue the stops were at

At Forty-second Street a spur curved off the Third Avenue mainline toward the domed splendor of Grand Central Station. The wedge-shaped depot with its delicate carpenter Gothic tracery, railings, cresting, and flanking platform pavilions suggests a Swiss chalet. A one-car shuttle train, waiting for the start signal from the conductor, blows off its safety valve in impatience. COURTESY THE NEW YORK HISTORICAL SOCIETY.

Ninth, Fourteenth, Eighteenth, Twenty-third, Twenty-seventh, Thirty-fourth, and Forty-second streets. At Forty-second Street a spur went west to connect with the new Grand Central Depot of the New York Central Railroad. After about a year of direct routing to the depot and when the main line of the Third Avenue el had been built farther north, shuttle trains only were run on the spur. This shuttle was continued until 1924, when the city took possession of the structure and removed it on the grounds that it was no longer needed.

In 1878 the main line on Third Avenue was continued north to 129th Street, including twelve new stations, the last being the northerly terminus. In the same year yards and offices were opened between 128th and 129th streets from Second to Third avenues, with a limited waterfront on the Harlem River for handling coal. Property located between 98th and 99th streets and Third and Fourth avenues was purchased in 1879 for repair shops and offices. By January 1, 1879, the two lines of the New York Elevated were carrying 84,000 passengers daily, and the population of New York City had grown to 1,164,673.

In March 1879 the Third Avenue line opened another branch built under its right to connect with railways, bridges, and ferries. The City Hall branch extended from the Bowery and New Bowery through Chatham Street (later Park Row) to Tryon Row and later connected with the west terminal of the Brooklyn Bridge, then under construction. After the opening of the bridge in

1883 this branch was extraordinarily busy, because trains operating over the bridge carrying heavily loaded commuter traffic from Brooklyn and Queens connected with Manhattan elevated trains at this point. The third spur on the Third Avenue line at Thirty-fourth Street provided a direct link with the East River ferries of the Long Island Railroad to Long Island City. Shortly after it was opened in 1880 the branch was a popular connection to the races at Belmont Park Race Track until 1905, when horse racing was prohibited. The spur was also used as testing ground for the elevated's experimental electric trains in the 1880s and 1890s. El service on the branch was discontinued by the company in 1930 and the structure later torn down.

Of the four elevated railroad lines on Manhattan, the last built was the one on Second Avenue, which was part of the original Gilbert routing. Since its path overlapped part of that granted to the New York El, an agreement was made between the two companies that, where the routes coincided, the cost of the structure and the expense of maintenance were to be divided in some fair way. Work on this line was begun in 1879 at the corner of Allen and Division streets. From Chatham Square the route followed Division to Allen then north on Allen and First Avenue with stations at Canal, Grand, Rivington, First, Eighth, Fourteenth, and Nineteenth. At Twenty-third the el turned west one block to Second Avenue then north for the rest of its route to Sixty-fifth Street. A storage yard for equipment was built at Sixty-sixth Street. The first experimental train ran over the section from South Ferry to Sixty-fifth on January 15, 1880, and the line was ready for regular service on March 1.

During the following months of 1880 founda-tions were laid and ironwork completed to the terminus at 129th Street. At this time, because of the difficulties of common routing between South Ferry and Chatham Square, Second Avenue el trains operated to South Ferry, and Third Avenue trains ran to the City Hall station via the City Hall branch. In 1882 an interlocking switch machine was installed at Chatham Square junction, allowing Third Avenue trains to operate to South Ferry without grade crossing. Because of the junction of the East Side lines at Chatham Square there was great passenger congestion at the station, passengers making changes to various points: Brooklyn Bridge trains, City Hall, South Ferry, uptown via the Third Avenue el, and uptown via the Second Avenue line. An overhead bridge was built connecting the north Chatham Square station (Third Avenue Line) and the South Chatham Square station (Second Avenue line), permitting free transfer of passengers between the two lines. Another very busy station was at the South Ferry interchange, where all the tracks of the East and West side elevated lines converged, enabling passengers to transfer to any elevated or to the ferries.

According to elevated railway authority William Reeves, the Second Avenue el was built in record time with few construction problems and hardly any legal or financial difficulties. Also, by the time of the construction of the final line the public was not so bitterly opposed to the roads as they had been earlier, perhaps because the el had brought about an increase in property values on the West Side and had clearly demonstrated its facility for quick transport. Since the West Side elevated was enjoying excellent corporate health and earning good public relations, it no longer seemed the absolute anathema that it once had.

6 The Financiers

The chronicle of the financing of New York's elevated-railroad system is a lusty story in itself—filled with greed, altruism, stormy stock manipulations, quick-profit speculation, merger, and eventually monopoly. Some of the most powerful robber barons in the nation invested in the elevated, recognizing the rich potential profit from efficient transportation in so populous a city. A yearning for control of the giant network eventually led to warfare among three of the most brilliant financiers—Cyrus Field, Russell Sage, and Jay Gould—which resulted in the total destruction of one man and virtual monopoly for another.

Unscrupulous manipulation of corporations flourished during the generation or so after the American Civil War, as individual enterprise was then seldom checked by government controls. Social Darwinism was the prevailing ethic, and the

Securities and Exchange Commission was many years in the future, as were antitrust laws and most interstate-commerce regulations. Plentiful capital was available for investment, and there was a large and cheap labor supply—largely immigrant and not much unionized. The mid-century American marketplace, especially New York City, was a ruthless exchange. There was little room for the weak or the foolish. In every level of business it was root hog or die.

Cyrus Field became interested in the adolescent New York Elevated Company in 1877. The line was then having some financial difficulties, and he offered his fortune and his services to the company if they would agree to his terms. Various reasons have been suggested for his interest. One Field biographer and close friend, Murat Halstead, believed that his motives came out of noble instincts and humanitarian concern since the elevated railroads, like the Atlantic Cable, would be of enormous public service by offering cheap transportation for the masses. Because Field already had money and success, Halstead concludes that he was obviously searching for good works. Another biographer, Richard O'Connor, argues that Field was just another nineteenth-century entrepreneur, wanting as much as others the wealth and position money could buy, uttering Christian pieties while the dollars rolled in. Resting behind his impeccable social position, said O'Connor, Field tended to cast his eyes downward in innocence while his confederates Gould and Sage did the actual dirty work of manipulating Manhattan Elevated stocks. Whatever his motives, there can be no doubt that Field's entry onto the scene did a great deal to bring prosperity and success to the elevated railroads, contributing as he did not only substantial capital investment for construction and expansion, but also his talents as promoter and organizer as well.

In 1877, at the age of fifty-six, Cyrus Field was at the peak of his fame—probably one of the most respected men in the Western world. He had done what many thought to be impossible; he had laid the cable connecting the New and the Old World, laying a copper wire 1,950 miles long through water two miles deep. His vision and endurances were remarkable, for during a nine-year struggle and after successive failures he continued to have faith in the project, going on to promote further attempts when most men would have admitted defeat. In 1866 the cable was finally laid successfully.

When asked by a reporter whom he thought to be the greatest living American, the pious Peter Cooper put his finger on a portrait of Cyrus Field saying, "Field did what no other man could. He saved the scheme and brought victory out of despair." And so it was that the son of a poor New England clergyman who had begun his career in New York as an errand boy and clerk at Stewart's Drygoods Store on Broadway was honored round the world. After his success in 1866 President Andrew Johnson had a gold medal struck, and the King of Italy conferred on him the cross of the Order of St. Maurice. He was given the Grand Prix at the Paris Exhibition, was often called Lord Cable, or Cyrus the Great, and compared with Christopher Columbus.

In 1877 Queen Victoria was proclaimed Empress of India, Commodore Cornelius Vanderbilt died at his home on Washington Square, and Cyrus Field became president of the New York Elevated. In May he had written to David Dows, prominent stockholder in the road, that he would be willing to purchase a controlling block of the stock in the debt-ridden line and serve as head without salary, but with provisions. He wanted the company out of debt. To achieve this, Field proposed that the stockholders accept bonds of sixty cents on the dollar for what was owed them. They agreed. Thereafter all future purchases were on a cash basis. As a further economy he canceled contracts for building the proposed Third Avenue line at a cost of $1,200,000 per mile as being excessive. Elevated railways were then ten times as expensive to construct as surface lines, and their average passenger fare much lower. The contracts were then negotiated at a lower figure and the line completed in slightly more than a year.

Field seemed delighted with his aerial railway, and his enthusiasm and the prestige associated with his name seemed to rejuvenate the whole operation of the company. Aware of the need for good public relations, Field was very skillful at promoting the advantages of elevated travel. That summer he organized a rally at Chickering Hall for friends and supporters of the el, hoping to encourage investors in the company and defeat the growing opposition to their planned, large-scale expansion. Some of the rhetoric that evening was irresistibly quotable, especially the florid speech of Charles O'Conor: "The cheap and rapid transportation on the elevated road . . .will give healthful and pleasant homes in rural territory to the toiling millions of our commericial and manufacturing centres. It will snatch their wives and children from tenement-house horrors, and, by promoting domesticity, greatly diminish the habits of intemperance and vice so liable to be forced upon. . .them by the present concomitants of their city life."

To curry favor with the New York press corps, Field gave in December a "Feast of Thanksgiving" at Delmonico's restaurant, attended by editors, elevated-railroad officials, champions of rapid transit, and potential investors. Simeon Church, crusader against the nasty horsecar and an el enthusiast, pronounced the principal oration after the heavy dinner, earning great waves of applause when he concluded that, "Hanging by

The elevated railroad station at Twenty-third and Sixth Avenue was one of the busiest and most elaborate in the entire system. Close by were many department stores and shops, theaters, and the Grand Opera House.

the strap, like the horrors of the Black Hole of Calcutta, shall now pass into history.'' Despite this skillful PR program, investments in the el remained rather cautious. Field and his friend and Gramercy Park neighbor Samuel Tilden, candidate for president in 1876 and now governor of the state, were alone heavily involved. As earlier with his cable projects, Field went to England for capital, since English money had backed many other American railroads. He believed that New York Elevated stock might be easily disposed of there at a tidy profit since the property was so well known, so obvious to the eye, and so clearly of enormous earning capacity. There is no record of how much of the stock he sold on that trip.

On his return Field became absorbed in his elevated-railroad projects, showing up at the construction site over on Third Avenue to inspect the ironwork, speaking soothingly to residents there who objected to the ugly appearance of the el in their street, and promoting in all circles the advantages of elevated travel. In time his efforts began to bear fruit, for early in 1879, less than two years since his heavy commitment to the com-

pany, the New York Elevated was for the first time showing a healthy profit. The price of the stock had increased ninefold, and the standard of service was much improved. In terms of public patronage the el was making great advances: when Field took over in 1877 the total number of passengers carried was 9,000,000. By 1879, when the Ninth Avenue line had been extended to 155th Street, the number had jumped to 46,000,000. By the time the Third Avenue line reached Sixty-seventh Street the passenger total amounted to nearly 61,000,000 for the year.

All seemed to be proceeding smoothly for Field until his close friend and trusted partner in the New York Elevated Company suddenly and for no apparent reason defected, selling all of his shares in the line and causing the stock to drop thirty-six points. Field was outraged by what he felt to be a betrayal of faith, saying in his most righteous tone, ''that my feeling for Mr. Tilden is one of intense disgust; that nothing, *nothing* would induce me to have anything more to do with him whatever, socially or in business.'' Tilden's sellout indirectly forced Field into association with two of the sharpest stock manipulators of all time—Jay Gould and Russell Sage. Both of these men were big investors in the Metropolitan Railway Company, operator of the Sixth and Second Avenue elevated lines. But while Field's road was very healthy, their elevated was shaky, though still a vigorous competitor. Sage suggested the idea of merging the two companies to eliminate competition, through the Manhattan Company, which had been established by the Rapid Transmit Commission of 1875. In this agreement the stronger company did not lease the weaker one; neither did one have any control over the other. Why Field should have permitted his prosperous New York El to enter into such a union with the less prosperous competitor is not clear. However, he agreed to the terms, perhaps out of desire for a single, coordinated transportation system, and the old Manhattan Company was converted into a holding company. Field's goal of creating a unified transit system was a worthy one, but in joining up with Gould and Sage he was putting himself in league with men he was ill-equipped to handle.

Russell Sage was master of the corporate merger and one of the shrewdest and most conservative of all the great financiers. Even as a young man in Troy, New York, he was known as a sharp trader. When but twenty-nine he was elected Alderman of Troy and later Treasurer of Rensselear County. He served two terms as a congressman in Washington but gave up politics because there was not enough money in it. By 1863 he moved to New York City to devote his total attention to stock and finance; there he allied himself with Jay Gould and made his fortune, mainly in railroad ventures. In those days the principal business of the country was railroads. During the last twenty-five years of his life Sage was best known as a moneylender; at one time he is said to have had $27,000,000 out on loan. In spite of his colossal fortune his frugality was famous. Plain food and cheap clothing satisfied him as well as the richest. He had little or no philanthropy in him. He often embarrased his wife (who did not share his parsimony) by his distressing habit of clutching a man's lapel, rubbing the fabric between his fingers, and asking how much had been paid for the suit. He was widely disliked for his miserliness, though after his death in 1906 his widow, Margaret, in a sort of wild irony, gave away a good part of his seventy-million-dollar estate, establishing the Russell Sage Foundation, "to improve social and living conditions in the United States." She was also a great benefactress in New York City, presenting huge sums to buy rhododendrons for Central Park and to restore the governor's room in City Hall.

Sage and Gould's alliance was long and comparatively harmonious. Neither man had any strongly developed talent for freindship, but common interest in money brought them more closely together than anyone outside their families. Though Sage was twenty years older than Gould, they apparently understood their inordinate desire for money and fiscal power and feared the distrust and hatred engendered by their success.

For a time after the Manhattan Company merger the consolidated railroads did well. The price of stock went up as brokers and the public invested with confidence. Now that they were in a way partners Field and Gould began to work together in other business ventures, such as the development of the Wabash Railroad. Field served as president of the road for a time, profiting handsomely through his alliance with Jay Gould. At the same time Field continued his interest in laying the Pacific cable, though there were a great many obstacles in his way. At one point he set out on a voyage across the ocean to study personally the difficulty of laying a cable there; yet the tremendous distances between islands seemed to discourage him. On his return to New York in mid-1881, he found Manhattan Elevated in poor financial shape. Jay Gould had been up to his old tricks in his absence creating a bearish atmosphere in which the elevated stocks would sink to a fraction of their level so that he could later buy it up at depressed prices and seize control.

For some time Gould had been watching the growth of the elevated railroads in the city, pondering the chance for a financial coup—his particular specialty. He had already acquired control of the Western Union Telegraph Company and also the New York *World*, which would be his chief weapon in waging his attack on Manhattan Railway. Assured of a cooperative press, he undertook to drive the stock of the els down. Gould was master of this technique of destroying a company in order to gain control of it. He first undermined faith in a corporation to depress the stock; then if that were not successful enough, he wrecked the company and bought it as distressed merchandise. At a mere hint from Gould his editor William Hurlbert at the *World* would oblige by

Jay Gould: the wizard of Wall Street and president of the Manhattan Railway Company.

publishing stories critical of the Manhattan Railway Company. Other journals picked up the *World's* stories that the company was on the rocks and circulated them all over the East Coast. The Philadelphia *North American* reported that Manhattan's stock was "worth fully the ragman's price for the paper on which it was printed, but hardly more." As a result its market value sank further.

Gould and his ally Sage were then ready for the *coup de mort*. They persuaded the attorney general of New York to declare Manhattan Railway insolvent, sue for revocation of its charter, and place it in the hands of receivers, all, of course, to be selected by Gould. At the same time irate stockholders filed suits against the company, alleging that they had been swindled; these stockholders were all friends of Gould. All the while Cyrus Field remained unsuspicious, though he quickly began court action to cancel Manhattan's lease on his New York Elevated Company. Gould's elaborately staged farce was working. By July 9, 1881, Manhattan stock closed at 23, and by the end of the month it had fallen to 15½, the cue for Gould to move in and begin buying secretly. While the stock rallied with his purchase orders, Gould continued publicly to do all he could to destroy the worth of the corporation. By October

he had bought up all the shares he needed for control—48,000—thus assuring himself a complete grasp of the rapid-transit facilities in the largest city in the country. Having gained control, he, Sage, and Field formed a triumvirate to operate the elevated lines, dividing the spoils neatly: Field as president of the New York Elevated, Russell Sage as president of the Metropolitan Railway, and Jay Gould as head of Manhattan.

Under the new management Manhattan stock recovered beautifully. The receivership was abruptly terminated, and the *World* quickly reversed its position and began publishing glowing articles about the worth of the road. The *New York Times*, on the other hand, took quite a different stand, commenting editorially on the disgraceful stock-jobbing scandal of Messrs. Gould, Field, and Sage. By November Manhattan was selling at 55, the same figure it sold for just one year before Gould began his raid. It had taken him but six months to undermine the city's elevated railroads, terrorize and drive out their stockholders, and gain control of their management.

With the reorganization of the Manhattan Company $26,000,000 worth of new stock was authorized, all going to Manhattan, Metropolitan, and New York Elevated stockholders. When the pie was divided, the three financiers feasted to surfeit. At this time Gould succeeded in doubling fares on el trains, raising the price of a ticket from five to ten cents. This hike in the cost of a ride caused a burden on the average workman, whose weekly paycheck in the 1880s came to little more than eight dollars. Public opinion and the *Times* vigorously opposed the greed of the increase, but everyone involved in the coup had reason to be delighted—everyone, that is, but Cyrus Field, who very much disliked the idea of having his name linked in the newspapers with the notorious Gould and Sage. Although he had profited handsomely from their handiwork, the Puritan prig in him made him fearful for his reputation.

Sage and Gould were quite accustomed to public criticism—even hatred—but Field, feeling himself a cut above these common moneymen, was so appalled by it that he began to have pangs of conscience. As his guilt worsened under the attacks in the *Times*, Field used his leverage in the company, which, together with public opinion, forced a fare cutback to five cents for all lines of the el system. Gould was, of course, furious over

Field's hypocrisy as well as his bad business head, but he kept his outrage to himself as was his manner. The mythology has it that Gould "ate his revenge cold," waiting for the time he could teach Field a lesson in loyalty.

Superficially, their friendship remained intact after the fare episode. Gould even went so far as to arrange for Field to buy up a large block of Western Union stock and supported him as a director. However, as much as the eminently respectable Cyrus Field may have disapproved of the Wizard of Wall Street, he could not deny that through his alliance with Gould he had grown a richer man by far. Field was now able to live in a grand style, both at his fashionable town house in Gramercy Park and at his country estate called Ardsley Park on the Hudson River between Dobbs Ferry and Irvington. There on the Hudson palisades he built a massive, mansarded and verandaed manor house surrounded by seven hundred acres of garden, family villas, and servants' cottages. Ardsley all but adjoined Lyndhurst, the estate of Jay Gould and his family, and was close by the country home of Russell Sage. North of Field's own mansion were homes he had built for his elder son, Edward Morse Field, Cyrus Junior, his son-in-law, Dan Lindley, and his nephew Frederick Stone. His two daughters, the unmarried Alice and the widowed Isabella, lived in the big house. Ardsley was considered one of the showplaces on the Hudson, a regular slowdown spot for tourist steamers that paddled up and down the river. In addition to his city and country seats, he owned a ninety-eight-foot steam cruiser, the *Inanada*, which he occasionally used for entertaining and for commuting to Manhattan.

In the city Field's holdings were equally impressive. In 1881 he purchased two newspapers, the *Daily Mail* and the *Express,* merging them into a single, two-cent daily called the *New York Mail and Express*. He then acquired one of the most desirable business locations on Manhattan, a site between the lower tip of Greenwich Street and Broadway on the Battery, where he put up the Field Building, a sort of protoskyscraper edifice towering three hundred feet over the pavement and providing on the observation deck and restaurant a splendid panorama of the city and harbor. *Leslies Illustrated Weekly* called it "the finest office structure in the city." Into the Field Building he moved the staff and presses of the *Mail and Express,* his son Dudley's law firm, Cyrus Field Junior's brokerage business, and Edward's commodity firm, thus securing a sort of feudal fiefdom both downtown and in the country.

Field's investments were also very great: he owned $3,000,000 in railway stock in nine different companies, other securities in coal, steel, Western mining, utilities, and 15,000 worth of shares in Edison's new Electric Light Company in London. He was a principal proprietor in the Newfoundland Land Company and a fifth owner of the Acadia Coal Company in Nova Scotia. He owned four entire city blocks of real estate in Manhattan, and then there were the 10,000 shares he held in the Manhattan Railway Company. Well able to afford it, Field financed a company (The United States Electric Railway Company) for his nephew Stephen to work on developing a satisfactory method of using electric motors to power elevated trains. Field did this on his own, since the rest of the directors in Manhattan Railway refused to consider such a damned foolish notion. From the vast accumulation of wealth it would appear that the Cable King belonged among the fraternity of men called the robber barons—though his sensibilities would tend to make him stand somewhat apart.

Field's association with Gould and Sage and his enormous financial success had not in any way disminished his prestige. He continued to cultivate close friends such as Dr. Adams, Henry Ward Beecher, Dean Stanley, Frederic Farrar, Archdeacon of Westminster, and Presidents Grant and Garfield.

It was in the mid-eighties that Field began to speculate very heavily in the stock market. Soon the fever spread to members of his family, and his sons began to buy large blocks on margin. Since the Manhattan Elevated stock had done so very well (increasing from earnings of $303,200 in 1877, when Field came in, to $7,000,000 in 1885), he was struck by the idea of getting control of the elevated roads and squeezing out Gould and Sage. Having a legal monopoly on New York City's rapid transit, Manhattan Elevated was to be sure a preeminent corporation to control, for in spite of a great decline in railway property in general, Manhattan had within four years advanced from 15¼ to 95½, or 500 percent. This advance was due to steady increase in earning capacity and patronage in the elevated system.

Through 1886, while Field's agents were scurrying around Wall Street buying up thousands of shares, Gould was baiting his trap and calling it sweet revenge for Field's lowering of the el fare six years before (remember that Gould was still a major stockholder in the Manhattan Company). Field's heavy buying had driven the price of the stock up to 128, then to 164, then further to 175, three times its value in 1881. Despite this inflation Field was determined to corner Manhattan Elevated at any price. Through the following months his holdings climbed from 10,000 to 70,000 shares, and he borrowed more money from banks as more Manhattan appeared on the market. Though warned by one of his brokers, T. W. Pearsall, of

the dangers of buying so much stock on credit, Field recklessly continued to buy. It was at this point, when Field was positively loaded with unpaid-for Manhattan stock, that Gould and Sage dumped their own vast holdings of the stock on the market. As fast as he could, Field bought it up at the inflated going price. The fierce pressure from buying began to weaken Field, who had to unload his Western Union shares to cover the new purchases. Incidentally, the value of Western Union shares had recently dropped because of Gould's manipulation. Cyrus Field was being caught in a terrific squeeze, yet he kept on with his wild buying raid—up to 80,000 shares.

And then it happened—on Friday, June 24, 1887. The bottom fell out; Manhattan stock dropped ten points, then crashed thirty-two more by noon. A crisis had occurred, and Wall Street began to panic. Quickly Gould withdrew his funds from banks supporting Field, and those houses responded by calling in Field's loans. Field was

Three of the principal financiers of the New York Elevated are represented in this 1882 lithograph, The Kings of Wall Street, *left to right: Cyrus Field, Russell Sage, Rufus Hatch, Jay Gould, Sidney Dillion, D. O. Mills, William Vanderbilt, August Belmont, George Ballou, and James Keene.* COURTESY THE SMITHSONIAN INSTITUTION.

trapped, overloaded with thousands of unpaid-for shares and collapsing credit. Though he must have guessed who his antagonists were, he had no alternative but to appeal to Gould for funds. Gould agreed to help him, and thus Field sold his 75,000 shares in Manhattan, bought mostly at the inflated price of 175, to Gould for 120 a share. In that single sale he lost $4,250,000. The balance of the stock, some 10,000 shares, he was forced to sell as low as $90, bringing the total loss for the day at close to $6,000,000, his total fortune.

Lord Cable was ruined. Almost everything was gone: his stock, his securities, all of his investments. Only the real-estate properties remained, and they only temporarily. As Samuel Carter so well commented on the debacle, "Perhaps in no other period was failure more intolerable." Success and wealth were so synonymous in the Gilded Age that a man who had lost them had slipped from grace, and Cyrus Field was reduced to charity. Gould's niece, Alice Northrop Snow, recalled in her biography of Helen Gould (daughter of Jay) an evening at Lyndhurst when Field, in desperate straits, called on Jay Gould to beg for help, "looking physically sick . . . a drowning man . . . a picture of abject despair." Field never recovered from the emotional holocaust of his failure. With the help of wealthy friends and by selling off bit by bit his remaining possessions he lived out his remaining years in modest circumstances until his death in 1892. He was buried quietly in his childhood home in the Berkshires.

Whether in fact Gould actually was the cause of Field's tragic ruination is a debatable point. Russell Sage positively denied it. Gould's family insisted that Jay had saved Field from bankruptcy by buying the Manhattan stock from him. John T. Terry, a broker who acted as Field's intermediary, told the New York *Tribune* that Gould had played a completely heroic role in the whole affair. Even the *New York Times,* an outspoken critic of Gould, said that his stepping in as he did saved several financial institutions from utter ruin. Most people then and for years later, however, insisted on putting the blame squarely on Gould, partly because he had become such a positive symbol of all the evils of capitalism. One is still tempted to concur with the majority sentiment, that the shrewd financier led the ingenuous Field into a speculative trap and then cornered and smashed him. Such a thesis may be true, though there is no

conclusive evidence available to prove it. Julius Grodinsky, Gould's most scholarly biographer, seems to conclude that, while Gould profited immensely from Cyrus Field's failure, he was not the primary cause of it.

Whatever the truth may be, it remains that now Gould had become the undisputed master of New York's rapid transit. And it was a very valuable piece of property. Gould valued it so highly that he made Manhattan Elevated a permanent part of his holdings, never giving it up as he usually did once he controlled a corporation or using it for further speculative purposes. So important was the elevated to him that he took an active part in its management, serving as president of the company until his death. He installed his oldest son, George, as vice-president of the company, further marking the property as exclusive Gould family territory. As the earning power of the els expanded through the late 1880s and 1890s, the corporation became one of the strongest and most lucrative in the country, much sought after by bankers and brokers as a safe investment. The stock was placed on a dividend basis, and the property proved to be, along with Western Union and the Missouri Pacific Railroad, the foundation of Gould's personal fortune in his lifetime and one of the principal pillars of his estate after his death.

Jay Gould was certainly one of the most brilliant of all of the nineteenth-century entrepreneurs, a figure whose cunning and ruthless energy matched the lustful business ethics of the time. Son of a poor New York Scots-English farmer, he went on to dominate the policies of corporations that controlled more railroad mileage than any other single person or group of persons in the country. According to Grodinsky, although Gould had neither the brilliance nor the cold-heartedness that he was popularly believed to have, he did possess a cold-blooded unscrupulousness that allowed him to take full advantage of the primitive nature of the art of corporate law and adapt to his own ends the low state of political morals prevailing at the time. None of his contemporaries ever quite approached his genius for trickery, his boldness and his talent for strategic betrayal and skill at stock manipulations, his bravado in looting a company and fleecing the stockholders.

Unlike Cyrus Field what Gould lacked most, probably, was a sense of public relations. Gould allowed himself to become the favorite whipping boy of crusading editors and reformers. He made a bad personal impression, and he did not very much care. He could afford enemies. He did, however, very much mind what the bad publicity did to his family, for in spite of his colossal wealth, neither he nor his family were recognized by New York society. They were never admitted into the charmed circle dominated by Vanderbilts and Astors, perhaps because of Jay's notoriety or perhaps because his fortune was considered "new money." If the Astor millions came out of a butcher shop, in part, and continued to gush forth from slum rentals, the passage of a few generations had sanctified it. In later years Gould was often called a Jew, but the family was indisputably Aryan, descending from the distinguished Gould family whose members had been military officers, deputy-governors, and chief justices of the Supreme Court of Connecticut. His great-grandfather, Colonel Abraham Gould, had been a revolutionary war commander. With all of his unbelievable wealth Jay Gould did not live very long to enjoy it. He died of tuberculosis the same year as Cyrus Field, leaving tens of millions to be squandered wantonly by his heirs.

7 Engineering and Construction

The engineering of the New York elevated rail-roads was a frequent topic of discussion after completion of the system in the early eighties. Although some people disliked the structure, most observers thought the elevated made a striking addition to the cityscape. The lacy ironwork construction seemed light and graceful, yet was surprisingly solid. Visitors to the city thought the el wonderful, being much taken by the very novelty of an aerial railway, especially the spectacular high section on the upper West side at 110th Street between Eighth and Ninth avenues, the grand serpentine curve, where the engineering was especially bold owing to its great elevation. The track was sixty-three feet above the street, supported on tall iron columns that seemed almost too frail to bear the burden of a heavily loaded steam locomotive and its train of cars. One contemporary tourist guidebook to New York City,

New York Illustrated, seemed to capture the Victorian enthusiasm visitors experienced when first seeing that thrilling high section of the el: "As one drives under this giant curved bridge and sees the trains gliding far overhead in the air, the imagination is fascinated by the daring of science in overcoming difficulties."

When new the el structure presented a very fine appearance, the ironwork freshly painted a light buff color. Overhead the new road reached along the great avenues of the metropolis like an immense arcade, its receding perspective vanishing to a point in the distance. It was a wonderfully new and extraordinary feature of street architecture, soon to become famous round the world. Like the imaginative ironwork of the Eiffel Tower or the train shed of old Penn Station, the el superstructure had intrinsic beauty, symbolic of scientific progress and technological achievement. Along

When new the Second Avenue el structure presented a fine appearance as the ironwork had just been freshly painted a light buff color. Overhead the new road reached along the great avenue like an immense arcade, its receding perspective vanishing to a point in the distance. COURTESY THE COLLECTION OF ROBERT M. VOGEL.

joined together. The New York Elevated at first favored the one-legged design by which the weight of the railroad was carried immediately over the columns that supported the structure. The Metropolitan Elevated, on the other hand, usually preferred putting the tracks down the middle of the street joined together by transverse girders.

Milton Courtright, chief engineer of the New York Elevated, explained that in the section of their East Side line in lower Manhattan along Front and Pearl the streets were so narrow that tracks had to be put over the middle of the street. Columns were set along the curb line and the tracks carried between on transverse spans. This sort of construction meant that the road underneath would be partially covered, but the sidewalks would remain clear and unobstructed overhead. In wide streets there were two possible locations for the tracks: either close together over the center of the street or independent, one on each side of the street. Along the Bowery and New Bowery, owing to the number of horsecar tracks, the New York Elevated put the columns along the

A train rounds the S curve at Coenties Slip on the lower East Side. It was still Little Old New York when this picture was taken in 1891, where the el meandered down through Pearl Street, past the Fulton Market, and the dock district to South Ferry. COURTESY THE COLLECTION OF ROBERT M. VOGEL.

with the Brooklyn Bridge the elevated railroads were for years a preeminent tourist attraction for the city, and for those who could not manage to see it first hand, stereoscopic views of it were sold by the thousands all over the land.

In its essential details the elevated of the great construction period of the late 1870s and early 1880s was the elevated of the 1940s and 1950s. Though the structure was altered and relocated from time to time, operations were extended farther into the suburbs, and third and fourth tracks were added to permit express operations, it remained in its engineering quite Victorian. Various styles of superstructure were used in erecting the elevated. Final choice depended on such factors as the chief engineer for the line, the contractor, the width of the street it would run through, and the character and complexion of the inhabitants there. No one company was consistent throughout a given line of elevated track: some stretches might have been one-legged, others of outrigger construction, independent track or curbs and ran the tracks in independent structures on each side of the street. But along the Third Avenue stretch engineers thought it best to move the tracks as far as possible from the houses since the upper stories were generally occupied as dwellings. Consequently, since the avenue was sixty feel wide, the two lines of columns were placed in the center of the street and connected at the top by graceful, elliptical arch girders.

Several contractors participated in the construction of the elevated network during the principal construction years since the work was too great a task for any single company. Among the most prominent general contractors in the second

generation were the New York Loan and Improvement; Clark, Reeves, and Company, of Phoenixville, Pennsylvania; the Passaic Rolling Mill Company of Passaic, New Jersey; the A. R. Whitney Company; the Edge Moor Iron Company, of Wilmington, Delaware; and the Keystone Bridge Company. The Company of Mills and Ambrose were foundation specialists only.

The actual work of putting together the elevated railroad superstructure obviously varied somewhat from year to year and from contractor to contractor; however, for purposes of simplicity in such a complicated and changing technology, the construction of a typical el trestle of the mid-1880s will be used as an example. The basic steps in el construction were: (1) making the preliminary survey, (2) building the foundations, (3) erecting the ironwork superstructure, (4) laying the track, and (5) painting and finishing.

One of the most famous stretches of the el was the breathtaking "Suicide Curve" at 110th Street on the Ninth Avenue line, where the engineering was particularly bold owing to its great elevation. C. 1900. COURTESY PRINTS AND PHOTOS DIVISION, LIBRARY OF CONGRESS.

An uptown section of the Third Avenue el in 1955—still a bold and forceful bit of civil engineering. PHOTO BY ROBERT M. VOGEL.

Engineers often ran into difficulties in their preliminary survey work of locating sites for posts, stations, and stairways, especially in the old, narrow, crowded streets downtown. In spacing and arranging the foundations for the columns it was not uncommon for the surveyors to encounter such impedimenta as sewer basins, hydrants, manholes, lampposts, gas and water mains, or storage vaults. Either these obstacles would have to be moved or the blueprints altered. Having plotted and located the foundation sites, engineers would determine the grade of the road and the length of the columns. As a general rule the el tracks were twenty feet over the street and level with the second-story windows of the houses by which they passed, depending on the elevation and depression of the street. In some sections of the city, however, the road was much higher. The land west of Central Park, for instance, is very hilly, and the trestle had to accommodate this geog-

raphy. Where the ground was low, the ironwork had to be quite high, some posts more than sixty feet high, and it became a gigantic feat of engineering to keep the track level. Such construction was also very expensive, calling as it did for towering masonry foundations resting on pilings driven forty feet into the ground beneath the ironwork. At another stretch of the el along the New Bowery the tracks ran along at the dizzy height of thirty feet from the street so that passengers were nearly forty feet high as they whirled along in their cars uptown or downtown.

The casual observer saw the el track as perfectly flat, but in its hills and hollows there were grades as heavy as three and a half percent. In fact certain grades were part of the basic engineering of the road. Whenever possible, stations were placed on summits in order to use gravity for starting and stopping trains. An upgrade was built into the track for a train to coast to a stop at a station; a

Like the imaginative ironwork of the Eiffel Tower or the old train shed at Pennsylvania Station, the el superstructure had an intrinsic beauty, symbolic of the nineteenth century's scientific progress and technological achievement. This scene is near 116th Street and Eighth Avenue, where the tracks were very high. COURTESY PRINTS AND PHOTOS DIVISION, LIBRARY OF CONGRESS.

The structures of the West Side elevated stretched rapidly uptown in the late 1870s, through a largely undeveloped part of the city. A construction derrick raises girders for the high section of the el at West Ninety-seventh Street and Ninth Avenue. Central Park is in the background. COURTESY THE NEW YORK HISTORICAL SOCIETY.

downgrade then helped it get moving on the restart.

During the foundation stage of construction workmen dug pits in the street about seven feet deep and nine feet square. If the soil formation was solid, a layer of concrete eight feet square and nine inches thick was first laid down; then on top were placed two large stones five inches thick and three by seven feet around as an anchorage for bolts. On top of the stones went a brick pier—four feet or more deep of about seven thousand bricks. (The piers could also be built entirely of concrete.) Piers of special design were sometimes required when engineers met difficult subterranean obstacles, such as sewers or water and gas pipes, or when the

ground was too soft to sustain the necessary weight. In putting up the Sixth Avenue line along West Broadway, engineers had to employ a good deal of ingenuity because of the weak nature of the subsoil there. To secure a solid foundation, workmen had to drive in piles twenty-five feet deep so that the foundation could be safely laid. After the foundation work was finished, the base casting holding the column was bolted on and the fender put in place. The base fender was a cast-iron device of bell shape designed to embrace the foot of the column just above the street level, protecting the column from being struck by the wheel hubs of passing vehicles. The space between the shell of the fender and the base casting was filled with cement.

The next stage in construction was erecting the columns and placing them in the base casting. This procedure was at first done by means of a derrick, wagon, and horsepower and later by a traveling derrick on the structure operated by steam power.

Along the Bowery, owing to the number of horsecar tracks in the street, the New York Elevated put their columns along the curbs and ran the tracks on independent structures on each side of the street. Later additional tracks were also built over the middle of the street.

Seven men and a team of horses could set up from ten to forty columns in a day. After they had been set on the base casting, columns were held temporarily in place with iron wedges inside the socket of the fender and the space between the rim and the column caulked with oakum to keep out water before cement was poured in. Once the horizontal girders were in place and the braces adjusted and riveted, the columns were carefully plumbed and held firmly by jackscrews until a cast iron cement was substituted for the oakum at the base. This cement was also used to fill hollow Phoenix columns to lessen corrosion and deaden noise. The cast-iron cement, which was thought to

be as strong as, if not stronger than, iron itself, was made by putting clean cast-iron turnings or borings in a solution of one ounce of salammoniac to one gallon of water. After being mixed, it was left to stand for half an hour to "warm up," then stirred with the cement. The mixture was then dropped in thin layers and thoroughly tamped. In forty-eight hours it became as hard as the iron.

Girder construction of the el was usually uniform on each line but specially fabricated members were sometimes required. Solid plate girders seemed to be preferred for all spans under sixty feet; for longer spans the lattice girder was usually employed.

After the ironworkers had completed their jobs, construction gangs moved in to lay the track foundations and rails. Crossties were laid ten inches apart, usually of yellow pine except on curves, crossings, and turnouts, where white oak, cap stringers, and shims were used. Crossties were fastened to the longitudinal girders by lag screws with washers at the bottom, the latter projecting under the top flanges of the girders, clasping the ties to them. On curves the outer rail was elevated about three inches. The T-rails, of four-foot-eight-and-a-half-inch gauge, were of Bessemer steel, from sixty to ninety pounds per yard. Fish plates, screw bolts, lag screws, clips, angle bar, blunt bolts, spikes, and nails were other fundamental tracking materials. Usual procedure called for every third tie on main lines to be laid twelve feet longer in order to provide a support for the plant sidewalk beside the track. On each side of the rails were placed heavy five-by-eight and five-by-ten-inch guard timbers bolted to the ties to prevent the locomotive and cars from falling to the street in case of derailment. The slot formed by the T-rail and the guard timber caught the wheels of locomotive and cars and held them on the trestle.

As a final step all the surfaces of the elevated structure were painted with high-quality material. One engineer recommended a metallic paint for first coat and white lead for the second. At first the hue was a buff color, but perhaps because of soiling from coal smoke, it was later changed to olive drab. All ironwork was thoroughly cleaned before painting, rough spots scraped and sanded, and those areas with new rivets or other raw metal primed. Seams and cracks in the ironwork and all joints, cracks, and sun checks in the timberwork were filled with linseed putty before the final coats of lead paint were applied.

The proper care of the vast elevated railway structure involved Herculean labor and great expense. Because the el network was ostensibly a great series of iron bridges, the potential for disaster from structural failure was frightening. As a result it was absolutely imperative that the thousands and thousands of vertical columns and horizontal girders have constant inspection and maintenance. The great number of lives depending on this constant vigilance demanded immediate repair of any defect discovered. The Engineering Department of the Manhattan Elevated was charged with all inspection and maintenance, though the Roadmaster's Department did the actual work, employing track supervisors, inspectors of girders and rivets, riveters, cleaners, painters, and carpenters. Each man had a specific duty on a specific division of the road; and the track was constantly patrolled, and rails, spikes, fish plates, and frogs were carefully scrutinized. Foundations and superstructure of the high sections of the el such as at the 110th Street curve were inspected every morning.

Each column was numbered so that any report of a defect could be quickly located by repairmen. The inspector's report would name the line, north or southbound track, between columns 6,150 and 6,151, such and such rivets loose. As soon as possible a riveting gang was dispatched to repair the flaw. When it was learned that a track was sagging or that a foundation had settled, a trestlework was erected beside the nearest column and the track lifted with jacks. If necessary a new foundation was built. When workmen were repairing the ironwork, the rivets were driven immediately after a train had passed to allow a few minutes for cooling before the strain of the next train.

One of the greatest obstacles in the construction of first-generation elevated railroads (Harvey and Gilbert) had been their enormous expense. Charles Fairchild estimated the average cost for a mile of double-track elevated to be $600,000, including a station for each track. According to Russell Sage, whose figure one would tend to trust, the total cost of building the four elevated lines when they were just completed about 1881 was $4,535,754.70. During the peak construction period in 1879 some six thousand workmen were employed on the roads. The cost of construction may have seemed high for the time, but any other plan for rapid transit such as the subway would have cost infinitely more. The elevated railroad was so practicable that its future success appeared assured, and the prediction of the backers came to be true as the income from the four lines grew each year as the number of passengers increased.

8 Steam and the Driven Wheel

Charles Harvey's idea of adapting the stationary steam engine to his cable-powered elevated railroad was brilliant, but sadly the method proved impracticable. After the failure of Harvey's company in 1871 the new directors asked for and were, surprisingly enough, granted permission by the state authorities to use small steam locomotives to pull two- and three-car trains over the road. Steam locomotives were, of course, the prime intercity passenger mover at that time, though for running through the city streets they had many disadvantages: they frightened horses; their exhaust was bothersome to pedestrians and residents; they befouled the air with their smoke and ashes; and because of the economics of steam boilers a locomotive small enough for street use was rather inefficient. Nevertheless, the state allowed the newly formed New York Elevated Railroad to substitute small locomotives for cable power,

perhaps because they wanted very much to see elevated transit continue in the city—air and noise pollution notwithstanding.

Because of the widespread objections to steam engines cavorting noisily through the streets of New York, the first-generation locomotives on the el were of the so-called dummy design. They were not actually dummies, of course: they were quite real and with genuine steam-locomotive innards. It was merely customary to call such small suburban or street locomotives "steam dummies" out of efforts to mute their exhausts and because they were concealed inside little decorative houses so they would not look like real engines—or so it was hoped. The first steam locomotives that ran on the New York Elevated were very quaintly camouflaged little vehicles with windows all around, and like the dollhouses they were, the exteriors were nicely lacquered, scrolled, striped, and gilded in

Though most of the el's 334 steam locomotives were sold shortly after the system was electrified in 1904, a few engines were kept on hand for construction work and repairs when there was no third rail. These were the last two Forneys, kept in storage by the company until 1942, when they were sold to a scrap dealer. Hopefully the sale was a patriotic gesture aimed at the war effort.

the prevailing taste of the time. The rich ornamentation was meant to make them look as little like the real thing as possible, perhaps even fooling the horses in the streets below into thinking they were just another horsecar running along some high-road. This excellent bit of applied psychology was obviously carefully planned to appease the enemies of the elevated, who had responded to the news of steam locomotives on the el by plastering the city with satiric broadsides showing how the overhead monsters would bring ruin to the city. As a result of their deceptive construction the little elevated engines were nicknamed "dummies," an appellation that lasted for el locomotives long after the cute wooden shrouds had been abandoned.

First-generation dummies were painted glossy black and bore distinctive names on their sides done up in fancy gold lettering on a wine-red background and surrounded by bright borders and fine stripes. Their names all derived from places in New York City and its environs, except for the original engine, *The Poineer: Manhattan, Battery, Greenwich, South Ferry, Yorkville, Harlem, Tremont, Spuyten Duyvel, Kingsbridge, Fordham, Morrisania, Yonkers, Westchester, Tarrytown,* *Williamsburg, Brooklyn,* and *Staten Island.* One engine built in 1876 departed from the norm by being named the *Centennial* to celebrate the anniversary of the nation; it was exhibited by the builder, the Baldwin Locomotive Works, at the Expo in Philadelphia. One staid New Yorker who staunchly opposed the use of steam engines on the elevated called the *Spuyten Duyvel* a "devil which belches fire and ashes on its evil way." With the expansion of the New York Elevated's lines all locomotives after 1877 dropped the individualizing proper names on their engines in favor of routine number identification, thereby loosing, as with digital dialing on the telephone, a good deal of personal associaton. Compared with the rich connotation of place-names, numbers are dreadfully lackluster.

Being rather experimental in nature, the first ten or so locomotives were designed by the road's superintendent, D. W. Wyman, and built to his specifications by local ironworks such as Handren and Ripley, the Washington Iron Works, William Harris Iron Company, Steel & Condit, and Hampson, Whitehill & Company. Three were built by the road's own shops. After 1875 the

The Metropolitan Elevated soon abandoned the dummy design on their locomotives, substituting engines with small, traditional cabs, exposing water tank and boiler. The bonnet-type spark arresters added an interesting frontal note. Some seventy of these engines ran in service until 1894, when they were superseded by the Forney.

engineering on the vehicles had been perfected to such a point that construction was handled by such builders as Brooks and Baldwin. The first-generation dummy locomotives were light four-wheeled engines with the cylinders placed inside the frames just in front of the forward axle and inclined so as to raise them high enough for the guides to clear. All had a wheelebase length of sixty inches, except for two (Nos. 14 and 15) built by Baldwin. The thirty-inch drivers were of the Moore patent, having cast-iron centers and steel tires with wood packing between. The engines were incredibly small, weighing about eight thousand pounds ready for work. In spite of their shortness, the *Railroad Gazette* claimed sanguinely that they rode smoothly with little pitching or oscillation.

On January 24, 1874, the *Gazette* carried an account of the elevated's operations, pointing out particularly the low cost of running the locomotives. At the time the company owned but four engines and ten cars, though more equipment was on the way. Coal consumption for one engine was said to be five hundred pounds of anthracite for eighty-two miles of running, or 6.1 pounds per mile. (El locomotives throughout the years of steam burned hard coal.) The method of taking on

coal was to fill an ordinary coal scuttle and carry it up to the engine; this supply was more than enough for a round trip of eight miles. One gallon of lubricating oil ran an engine for a week, or about 492 miles. The *Gazette* did not know quite what to make of the miniature scale of the elevated operations, commenting wryly that, "a visit to their dimunative shop opposite the Bowling Green is quite amusing, as everything is on such a small scale as to seem like playing at railroading."

When Cyrus Field as the new president of the New York Elevated embarked on an ambitious expansion program, the question of the best kind of locomotive to use on the new lines became very important. The light dummy engines that had been in use since 1871 lacked the tractive power needed for the new service; however, the success of the earlier engines caused some men to favor the old type. Since the new operations would demand heavier trains and high speeds, the choice of locomotive design came up for full discussion in the company before orders for new rolling stock were sent out. Feeling was divided between heavier hour-wheeled engines and an eight-wheeled tank engine of the Forney plan. When the advice of experts was sought, they were also about equally divided. As a result the board of directors decided to give out an order for twenty engines, half of them to have four wheels and the other half eight. A later engine order was divided the same way, but the next purchase for twenty-five was all for eight-wheeled engines.

The heavier four-wheeled locomotives built for the elevated after 1877 adandoned the masquerade covering and were twice as heavy as the first-generation engines. They carried a small traditional locomotive cab, heavily cantilevered at the rear, but retained the attractive decorative finish as in the old style. The small cabs somewhat resembled the dummy shrouds of the first-generation engines in design, having gracefully arched roofs and arched panels on the sides. Brass fittings and russia-iron boiler jackets sparkled brightly. And saddle tank, dome, cylinders, and wheel hubs and spokes were all nicely scrolled and striped to make the little locomotives very gay and visually pleasing. New York Elevated bought seventy-two engines of this type from Baldwin and Rhode Island.

Since the Gilbert-Metropolitan line on Sixth Avenue was completed before the extensions of the New York Elevated, they were obliged to decide on their own the best locomotive design to adopt. Their decision led to a lively debate among the officers of the company and various locomotive engineers. The principal question was whether they should adopt a four-wheel engine similar to those used on their competitor's road or to have a truck at one or both ends. Because the new road called for curves of as much as ninety-foot radius, the proper choice of wheel arrangement for their engines was a critical question. Some engineers objected to the dangerous rocking motion on the short four-wheeled locomotives, so the directors decided to eliminate that unsteadiness by using guiding wheels, or bogies, on the engine front and rear. These guiding wheels could swivel and allow the engine to negotiate sharp curves. Later it was found that only one truck was required for stability, and the wheel design was changed from eight-wheel to six-wheel. The first locomotives built for the Gilbert-Metropolitan el were a set of twenty-eight dummy engines delivered in 1878 by the Grant Locomotive Works of Paterson, New Jersey. Three times as large as the first dummies on the New York Elevated, the Gilbert engines weighed about fifteen tons. The entire locomotive was covered with a wood cab, camouflaging engine, boiler, and accessories; only

Dressed in their traditional livery of dark-red cab and gunmetal-gray boiler jacket, the trim little Forneys pulled thousands of trains over the elevated every month. Engine number 60 was an early locomotive of the type, built by Baldwin in 1878.

Basically orthodox in their engineering, the Forneys had no unusual features except for their arrangement: the weight of the boiler was carried over the driving wheels for maximum traction, and the water and coal supply were over a four-wheeled truck at the rear. With the built-in tender they could run forward or backward anywhere on the system.

the stack and safety valve protruded above the roof. The fancy cabs were built by the Pullman Palace Car Company of Detroit, Michigan, and sent to the Paterson Works for installation. George Pullman, incidentally, was at the time a director of the Metropolitan Elevated. The locomotive shrouding was designed and painted to look as much like the road's new apple-green Pullman cars as possible, so that the whole ensemble had a harmony of color and outline. The company directors had no doubt picked up the dummy engine idea from the New York Elevated; however, the style was a little late in coming and not entirely successful. The clerestory roofs on the Metro locomotives made them look just like ordinary horsecars, and the high skirts exposed too much running gear, adding further to a stumpy, top-heavy feeling. And since they were a good deal larger than the dummies on the other road, the Gilbert engines lacked good proportion and light-

ness. The New York Elevated had already given up their dummies because they were really not justified on el service, being so far removed from the street as to be an unnecessary conceit.

The Gilbert dummies were built with all the latest mechanical improvements, each being equipped with a steam brake; however they were found to have inadequate boiler capacity and could haul but three loaded cars. Later engines were made large enough to pull five. On subsequent orders the Metropolitan line, like the New York El, abandoned the dummy shrouding and substituted engines with small traditional cabs, exposing the water tank and boiler. The new engines were much trimmer in appearance than the cumbersome, boxy predecessors, and various types of bonnet-type spark arrestors added an interesting touch up front. The color scheme remained apple-green with dark-green and maroon trim. Some seventy of these second-generation 2-4-2's were built for the Metro by Grant, Danforth, and Rhode Island from 1878 to 1881, when the Forney design came into widespread use on the els. The original locomotives of the Metropolitan elevated continued in service long after Manhattan began operating the roads in 1879. Some of the engines were modified by removing the forward truck, because the wheel arrangement was thought too stiff to allow the locomotive to go around curves easily. Also, the dummy engines were rebuilt with traditional cabs. These engines continued in service until 1894 when they were taken off in favor of the standard Forney engine.

For ten years after Harvey's cable-operated elevated railroad failed, various engineers and designers worked to supply the sort of locomotive that would be best suited to haul trains on an elevated railroad track. After much experimentation it was left to Matthias N. Forney to come up with a design that would be so successful as to dominate the elevated locomotive scene right up until the time of electrification at the turn of the century. Born in Hanover, Pennsylvania in 1825 and educated in Baltimore, Maryland, he entered the locomotive building shop of Ross Winans in Baltimore, where he worked for two years as shopman and later as draftsman. From there

The popularity of the Forney locomotive was rather short-lived. Most of the engines disappeared quickly after the elevated was electrified about 1904. However, this veteran Forney was still running in 1930, hauling dead electric motor cars about the company's shops.

Forney moved to the Baltimore and Ohio Railroad's Mt. Clare shop, where he stayed in the drawing office until 1859. After an unsatisfactory interim in the commerical world, he decided to return to his railroad career, taking a position as draftsman with the Illinois Central Railroad in Chicago. While he was there, his fertile mind began working on what would later be known around the world as the Forney locomotive. In the mid-1860s there was a great need for small urban and suburban engines. Cities like New York were bursting at the seams, and the railroad seemed the only possible safety valve.

Forney patented his peculiar style of locomotive in 1866, intending it for railroads that ran frequent service but had light traffic. However, none was built until the elevated started to experiment with the Forney design in the late 1870s: in short, the scheme had been ignored for more than a decade. Once the elevated found the locomotive to be a good idea, the Forney type gained considerable attention and caught on rapidly elsewhere. Part of this rapid success stemmed from the engine's inherent character and its suitability to el passenger use, though Forney had never really intended his patent for such traffic. It is only fair to suggest that one of the reasons his design was received so favorably on the el was because of his standing in the New York engineering community and his national influence with locomotive builders.

The Forney was an 0-4-4 tank engine, that is one without a separate tender; locomotive and tender were one unit. Forneys were elementary locomotives, small, light, and simple—not sophisticated mechanically. Basically orthodox in their engineering, they had no unusual features except for their arrangement: the weight of the boiler was carried over the driving wheels for maximum adhesion, and the water tank and coal supply over a four-wheeled truck at the rear of the locomotive. All earlier engines run on the el had been of the saddle-tank type. Engines of Forney's design could be built to almost any size and weight; however, they were invariably produced for very light service and saw little mainline use.

Lack of a cowcatcher on the elevated Forneys gave them a somewhat strange appearance for an American locomotive of the period—marking them as distinctively elevated railroad stock. Actually, they did not need prows, because they ran on

The first locomotives built for the Gilbert-Metropolitan Elevated were from the Grant Locomotive Works of Paterson, New Jersey. The fancy dummy cars were furnished by the Pullman Palace Car Company of Detroit, Michigan, to match the road's new apple-green passenger cars.

an absolutely private right-of-way totally removed from the street obstructions that had created the need for such appliances in the first place.

As built for the elevated roads of New York, the size of the Forney was strictly limited, for no matter what the traffic requirements might be, locomotives of more than twenty-four tons were out of the question because of the structure—a relatively light, weak trestle. Even though the management realized that longer trains could move the burgeoning passenger traffic more efficiently, the idea was quite impossible on the el. Though mainline railroads could solve traffic congestion by putting on heavier locomotives, the elevated was hemmed in by strict structural limitations. Even at the peak of its development in the mid-1890s the Forney was a very small steam locomotive by mainline standards. Whereas passenger locomotives of that period were built up to forty tons and freight to a hundred tons, the el Forneys remained as light industrial engines. With this fixed weight in mind the design and manufac-

ture of these engines became highly refined in an effort to achieve maximum power. Some Forneys developed 140 horsepower. The selection of materials and the design of each part became a matter of careful consideration. Wrought-iron forgings were substituted for cheaper but heavier cast-iron parts. Lightness and thinness were constantly sought by construction engineers to achieve weight savings. The Forney was a little locomotive, and there was not much they could do about it.

The Forney proved ideal for the elevated, and in time the design was recognized as the perfect motive power for the roads, though not until after about fifty engines of other designs had been purchased for the service. After being first adopted about 1881 by the elevated system, the Forneys lived up to every expectation of their inventor. Designated as the four-coupled tank back with swivel truck, they showed such superior hauling capacity, tracking ability, and reduced cost of maintenance that they became standard power for all the Manhattan Railroad operations. With the built-in tender at the rear they could run forward or backward up and down the island. Once the Forney had been established as the best type of locomotive for the el, the chief engineer of the company determined to rationalize them by adopting standard design about 1885. A series of six classes from A to K evolved over the years, and all manufacturers were expected to build to these standards. All of the big locomotive builders had a go at production. Baldwin, Rogers, Brooks, Rome, Pittsburgh, Rhode Island, Danforth and Cooke, and Grant turned them out in large numbers. Dressed in their traditional livery of dark-red cab and tender and gunmetal engine jacket, the trim little engines pulled thousands of trains over the elevated thoroughfare every month.

The success of the Forney did not end on the Manhattan El. Brooklyn and Chicago adopted them for their elevated roads as well, and eventu-

ally there were 500 in el service altogether. Some mainline railroads began running Forneys for suburban service. The New Haven ordered a class of large 81,500-pound Forneys with forty-nine-inch drivers, and the Illinois Central built a batch of supertanker engines for their Chicago suburban service. The New York and Northern ordered 2-4-4 Forneys from Rogers and put them to work between Yonkers and their el connection at 155th Street and Eighth Avenue. The Cooke works built some very handsome little Forneys for the Staten Island Rapid Transit, and out West the Chicago, Burlington, & Quincy and the Rock Island Railroads joined the ranks of satisfied Forney customers. Apart from minor incursions the Forney dominated the el for twenty years before electrification. One rare exception in motive power was tried out in 1896 when an experimental

After the failure of Harvey's cable propulsion, the New York Elevated directors substituted small steam locomotives. The first engines were camouflaged inside decorative little houses to prevent scaring horses. The Spuyten Duyvel *was built in 1875 by Brooks.*

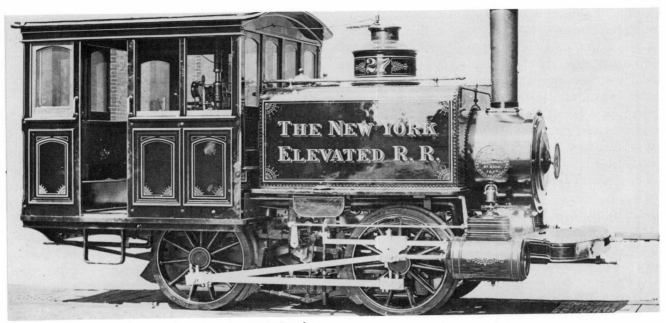

Second-generation locomotives on the NYER aban-doned the dummy shrouds. They carried small, tradi-tional cabs with gracefully arched roofs and decorative panels on the sides. This engine was built in 1878 by Baldwin.

compressed-air locomotive ran for a time on the Third Avenue line. The Hardie Pneumatic Motor was thought to be worth a try since it was silent, clean, and started more readily than steam, but after a thorough testing the exotic notion quietly faded away into the air.

After perfecting his locomotive design, Mat-thias Forney went on to a distinguished career. In 1870 his interest took a turn from mechanical engineering to journalism when he went to Chicago to assume the post of associate editor of the *Railroad Gazette.* Not long after the *Gazette* had moved their offices to New York because of the great Chicago fire of 1871, Forney and his editor-in-chief, H. W. Dunning, each bought a half interest in the magazine. In 1874 he published his *Catechism of the Locomotive,* which was widely read, and no doubt its prestige and weight helped the cause of the Forney engine to catch on. In 1883 he sold his interest in the *Gazette* and prepared to settle down and retire, but after three years of tiresome leisure he bought the oldest railroad magazine in the country, the *American Railroad Journal.* As editor and publisher he once again found himself as spokesman on national railroad and engineering matters.

Though the Forneys were run in almost constant service and were worked very hard on the el, the rugged little engines held up beautifully. Part of their success must in fairness go to the Manhattan Elevated's vigilant program of inspection and maintenance. The extravagant, continuous ser-vice of elevated rolling stock was in part created by a lack of terminal facilities, so that a good number of the engines were kept running twenty-four hours a day (with three crews) for a month at a time. Trains finishing a trip had immediately to start on another because there was no track for them to stand by on. Consequently, a careful program of inspection and maintenance was re-quired, not by lack of rolling stock but because of lack of room to keep it in. Because the strain of frequent starting, rapid acceleration, and sharp stopping resulted in loosening of screws and bolts, each engine went through a careful overhauling once a month, especially boilers, wheels, spark arresters, and brakes; boilers were washed out every month. Perhaps as a result of such attention there was never a boiler explosion during the el's steam years, though oddly enough the boilers of elevated locomotives were never officially inspected—apparently through a pecularity of law. After two years of use each engine was taken to the shop and stripped for thorough overhaul in the Manhattan Shops, a site covering two city blocks, one of the largest machine shops in the city.

The popularity of the Forney locomotive was rather short-lived since with the coming of electric traction small urban and suburban steam engines quickly disappeared from the roster of most roads. The Forney's biggest customer, the New York elevated railroads, were no exception. In 1903 third rail and electric cars took the place of 334 Forneys, most of which were in first-class condition. After standing unused and dejected in the Harlem yards, they began to disappear by ones, twos, and threes and soon their short exhausts were sounding again, not against the mortar and brick walls of New York City, but against the mighty trunks of California redwoods and Georgia long-leafed pines. Vermont marble quarries resounded with former elevated engines, and their notes ricochetted from rubber trees in plantations of far off Indochina. Some went to the uttermost reaches of the world—Alaska, China, Japan, Africa, and Burma—jerking flatcars and boxcars on logging roads and mining camps. They were said to have replaced the elephant in India. The red paint on the tenders and cabs and the gilt lettering and fine striping faded rapidly in the tropic sunlight since the new owners had no eye for such effete embellishments. Customers by and large were quarries, lumberyards, and contractors for construction projects, because the el Forneys were essentially light industrial locomotives, well made and with a lot of service left in them. And they could be bought cheaply secondhand.

Through the 1920s and 1930s a few old Forneys were used on elevated and subway construction work when there was no third rail, bringing ties and gear to the job. The last record of any of the little engines in use on the el was during the Second World War in 1942, when John Scharle wrote this postcard to the well known railroad locomotive enthusiast Charles B. Chaney:

In case your interested they took the last two "Forneys" out of storage and removed the canvass as they have been sold to a scrap dealer. They can be seen from the highway bridge walk at the 133 St. Yard of the 3rd Ave. El near the 133rd St. Station of the 3rd Ave. El. Cabs and tanks are painted red and striped one is #137 built in 1899; the other is #295 built in 1894. There is a strip plate on the steam chest with the name "Manhattan" on it. They still have the oil destination and back up headlights on the roofs. As near as I can learn 1933 was the last they were used. This was the hard winter and the 3rd rail was sleeted bad; they were fired up and used to pull stalled trains from the express middle track around midnight.

Skirting along the coast of Bohemia—a Forney locomotive and its train of three wooden coaches of the Third Avenue Line.

87

9 The Clattering Cars

The elevated came of age about the time the American railroad car was reaching its fullest flowering. First-class carriages were then called palace cars, and the name Pullman conjured up all sorts of rich associations of glamour and luxurious furnishings: Axminster carpets, architectural paneling, upholstery of deep cut-velvet plush or velour, fabrics tufted, roached, and pleated, and endless portieres, valances, fringe, and tassels. This was the era of the grand conveyance, not the plain utilitarian day coach. Period designs ran through entire cars in Chinese and English Chippendale, Byzantine, Baronial, Renaissance, and Spanish Mission motifs. It is no wonder that this prevailing taste in elegant railroad-car decoration was copied by the directors of the elevated roads in New York City, who wished just as much as the men who ran other passenger lines to offer exclusive, first-class service in mansions on rails. The men in control of the two elevated railroads during the time the system was being completed in the late seventies were shrewd businessmen, well aware of the importance of promoting their expensive, new, citywide service. Messrs, Field, Sage, Pullman, and Gould were experienced railroad men who realized that elevated travel was a new experience for many New Yorkers who had never used it before. Consequently, they seized upon the idea of offering a new and different service to the public, who were long tired of slow, depressing street conveyances. The management wanted to offer steam-railroad standards of elegance and service on the newly expanded and equipped lines, not just basic transportation as on the horsecar, but a choice of classes: a standard car well furnished and comfortable for five cents or an extra-fare parlor or drawing-room car with elegance and exclusiveness for ten cents. Since there

To promote their new service, which began in 1878, the Sixth Avenue elevated offered their patrons luxurious accommodations on extra-fare drawing-room cars. The car bodies, built by the Pullman Palace Car Company, were painted a fresh apple-green color with accents of bluish gray. Window areas were very generous, making the car light and cheerful. Inside, the furnishings were equally rich: oak and mahogany paneled walls and ceilings, architectural window cornices, leather seats, gilt chandeliers, and Axminster carpeting—an unheard of luxury for New York public transit. COURTESY THE MUSEUM OF THE CITY OF NEW YORK.

was competition on the els for the first time, with two companies in the field, the Metropolitan Road and the older New York Elevated Railroad, each sought the limelight by seeing who could outdo the other in having the most lavish coaches.

When the new rolling stock began to run along the recently completed Manhattan system, the press ran flattering accounts of the cars, calling them a new highmark in city transit. The New York *Herald* reported on June 8, 1878 that "The ladies were evidently charmed with the El road, its spacious and elegant coaches, decorated with all the taste and finish of a boudoir; its handsome young conductors in smart-fitting bright uniforms, . . . its breezy ventilation so welcome on a sultry day, and above all its rapid transit so exhilarating

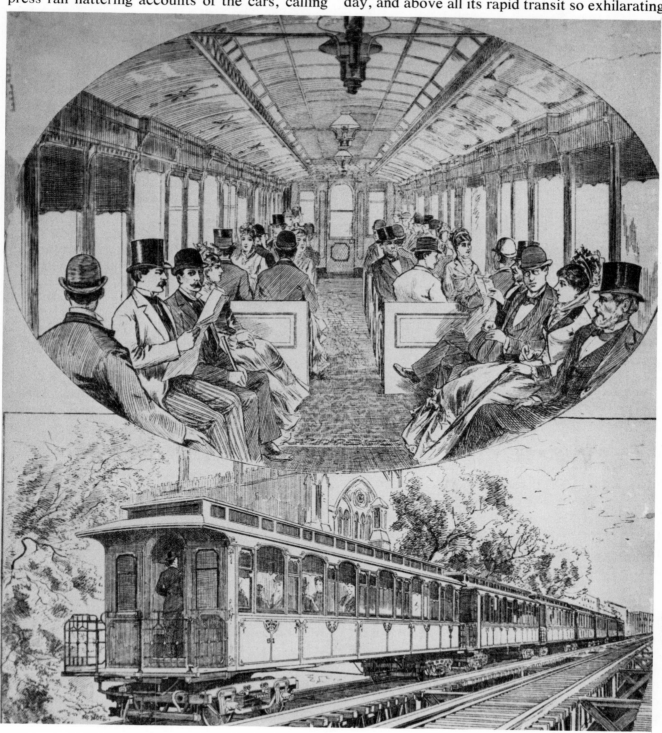

A typical elevated railway car, c. 1925. The center nook of cross seats and the side seats at the ends had been a standard arrangement for many years. COURTESY THE MUSEUM OF THE CITY OF NEW YORK.

to the spirits and gratifying to the mind.''

Of the cars built for two roads, those of the Metropolitan Road were probably the more elegant, because it was a new company and had to work harder to attract a share of the business for their lines on Sixth and Second avenues. When the first batch of new drawing-room cars was delivered to them by the Pullman Palace Car Company of Detroit, they presented a very fine appearance in their fresh livery of paint and varnish. One reporter remarked on how well they looked against the superstructure newly painted in a light buff color. Car bodies and trucks were painted an apple green with accents of bluish gray on the mouldings, window frames, and battens. Chamfered edges were rich maroon and the striping and Eastlake fancywork a deep shade of green and

gold. The window arrangement on the cars was thought to be rather special, being of the then fashionable two-and-one design: seven large, arched windows flanked on both sides by smaller ones. The whole car was unusually light and cheerful, excellent for looking out, and no doubt very pleasant to ride in. As originally delivered by Pullman the cars had monitor roofs, but the ends were rebuilt into the more fashionable bull-nose design about 1880, since the original style was thought to be rather old-fashioned. A traditional clerestory extended through the length of the car. At each end of the car were open platforms covered by projecting roofs and enclosed by ornamental iron railings and gates that were controlled by a spring beneath the floor in such a way that a brakeman at one end could open them both at the same time. The running gear consisted of two four-wheeled trucks with paper-center wheels; truck frames of wood resting on six elliptic springs and two double-spiral springs on each equalizing lever.

Inside, the cars were equally handsome. Rather than being covered with the ordinary canvas head-lining, so long in use, the ceilings were finished with oak paneling. The walls also were done in oak and mahogany and decorated in what was called "modern Gothic" or "Queen Anne style" (contemporary synonyms for the Eastlake style) with fresco decorations of flowers, plants, and arabesques to soften the harsh effects of the mechanical age. Instead of ordinary slat blinds Pullman used tapestry curtains at the windows hung on spring rollers concealed by cornices. The curtains were neatly trimmed in red leather to match the seats. Another well-advertised feature of the drawing-room cars was the Axminster carpeting, an unheard-of luxury in public travel for New York. The cars were quite spacious, having room for about fifty passengers with space for half as many more to stand. Red leather seats ran crosswise in the center of the car and lengthwise at each end to give more room for getting on and off. This arrangement of seats became standard on cars throughout el operations. The cars were heated by steam heat from pipes running under the

seats; at night they were lighted by three gilt oil-lamp chandeliers.

Drawing-room cars on the New York Elevated, while perhaps not quite so daintily furnished as those on the Metropolitan Road, did, according to the *National Car-Builder* of September 1878, present a fine appearance from their perch on the high track. Except for minor decorative details the New York El cars, first class and standard, resembled those on the other line. Owing to the difficulty with semantics in describing colors, the exterior shade of the car was described by various observers as deep red, carmine, lake, maroon, dark claret, and earth brown. Old-timers remember the color as very much like the Tuscan red used on the Pennsylvania Railroad cars. At any rate the basic NYER livery was some sort of rich, deep red with gold lettering and black striping. The interior was finished in black walnut and birdseye maple paneling with mahogany moldings and bronze trim. Ceilings were done up in the fancy Eastlake style. One difference from the Metropolitan cars was the seating, which ran in uniform rows down the sides as on a streetcar without the center nook of cross seats. The uniform side seats were characteristic of New York Elevated service before the Manhattan took over operations. Such cars were built in great number by Gilbert and Bush of Troy, New York, and the Wason Car company of Brightwood, Massachusetts.

When it appeared that the coaches were not going to fall off the overhead tracks as some people

Jackson & Sharp's demonstrators were rather unusual on the exterior. Instead of the conventional openend platforms, they had side doors and hemispheric bay ends, rounded like streamliners. The color was standard NYER claret red with gold striping and lettering.
COURTESY DIVISION OF HISTORICAL AND CULTURAL AFFAIRS, STATE OF DELAWARE.

The open summer cars that ran on the el between 1902 and the First World War were very popular with New Yorkers. They had airy park benches for seats, gay striped awnings, bright brass trim, and little sliding Dutch doors. COURTESY DIVISION OF HISTORICAL AND CULTURAL AFFAIRS, STATE OF DELAWARE.

had feared, the New York Elevated's old shad-belly cars were rebuilt along more conventional lines. Though an interesting feature of early el operations, the curious drop-center coaches did not last very long, apparently because they were considered dated and old-fashioned by the time the company began its expansion and publicity campaign in the late 1870s.

Anticipating lucrative car orders from the els' planned citywide service, the Jackson & Sharp car builders of Wilmington, Delaware, sent two fancy parlor cars to the New York Elevated as demonstrators in 1877, and while the design was never adopted by the rather conservative road, they are worth a few words because of some unique features. Instead of conventional open-end platforms the demos had hemispheric bay ends rounded like proto streamliners. Passengers entered through slide doors rather than the normal end entrances. Because the center-door arrangement took up the space of several seats, the builders installed movable benches that slid under the permanent seats. Since the doors of only one side were used in running up the line and the opposite in coming down, the sliding bench could be rolled out on one side or the other. As with other elevated cars of the expansion period the interior was lavishly paneled and carpeted, the only departure being the blue stained glass in the celestory and the lining material on the ceiling, which the *Railroad Gazette* thought to be wonderful innovations. Said the *Gazette,* "In appearance,

it [the ceiling] conformed to good taste very much better than the atrocious chromo styles of head linings which are now so universally used."

Out of efforts to prove that elevated railroad travel could be every bit as fine as mainline, the company directors had gone to a good deal of trouble and expense to provide plush extra-fare cars to transport the Wall Street elite. Rather soon, however, the idea of running two classes of service, deluxe parlor cars for the gents and ladies and less showy standard coaches for the general public proved a failure, probably because the travel time was so brief that few people, even wealthy patrons, saw little reason to pay the extra fare for a ten- or fifteen-minute ride. Or at least there were not enough people to make such service pay. In long-distance travel, where passengers were faced with long, tedious days on the road, first-class service could make a good deal of difference, but not for communting to work a few miles away. The scheme was no doubt an excellent psychological ploy to gain attention for the elevated's new citywide service from magazines and newspapers, but before long the companies abandoned the two-fare system and began charging a straight five-cent rate. The snobbery that had tainted the el for a while disappeared, and once again democracy reigned—bankers and dishwashers rubbing elbows in the same car on the way to work.

After finding themselves wrong in thinking that something special was needed in the way of rolling stock for the el, the directors decided to accept the standard wooden railway coach of the period as their model. The reasons for their decision are unrecorded, but it might be explained that because the mechanical operating officials of the elevated lines were regular railway men with many years of experience with steam-railroad operations, they would quite normally be familar with the standard style of coach, which had eight wheels, open end platforms, end entrances, center aisle, open body without compartments, and clerestory roof. Another reason for their adopting the standard coach rather than developing some sort of specialized car of their own was that the suppliers of the el were essentially manufacturers of standard railroad cars. Their designers and shop superintendents were convinced that the conventional design was the best possible for a railroad car, and if their advice was sought, as it no doubt was, they would naturally recommend the proved standard car as best and cheapest. As a result, with minor exceptions of seating and size, the elevated cars throughout their history remained miniature versions of the standard American railway coach, imitating their basic features although slightly reduced in size in an effort to cut weight and to be within the requirements dictated by the structure. El cars had about two-thirds of their seating put along the sides to create as much room for standing as possible and to create an open hallway at the ends to facilitate passengers getting on and off,

whereas the standard railroad coach had all cross seats.

Aside from color variations, cars on the elevated lines looked pretty much the same from the late 1870s right up to the end, since once the basic design was established there was no reason to change it. Unlike mainline railroads the el had but one type of service—short haul, frequent stop. With minor differences the cars through their history were characterized by batten-board side paneling, arched or pointed windows in a variety of arrangements, open platforms with iron gates, clerestory roofs, truss rods for bracing beneath the floors, and wooden construction. The cars were quite light, weighing from eight to eleven tons, while regular railroad passenger cars were from fifteen to twenty tons. Sturdy and well made, many el cars were rebuilt after electrification at the turn of the century and continued in active service for sixty years.

Some critics have argued that the standard elevated passenger car was not really a very good plan, saying that for a road that made so many frequent stops the narrow end entrances were points of dreadful congestion. The end platforms were awkward as well, since passengers had first to step onto the platform, then into the car, rather than entering the car directly. The open platform served little purpose, they said, because passengers could not use them to pass safely from a crowded car into another if they wanted to. These

Throughout their history New York's elevated cars remained miniature versions of the standard American railway coach, with eight wheels, open end platforms, batten-board side paneling, clerestory roofs, and open-plan interior without compartments.

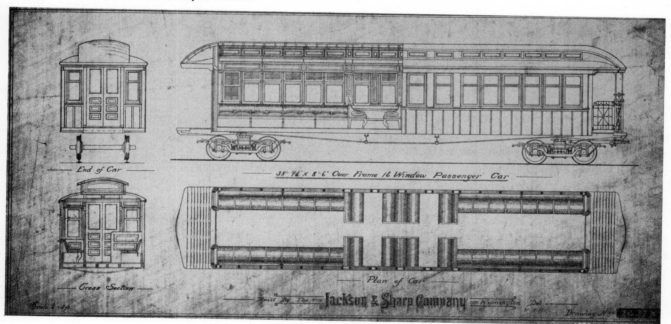

A standard elevated car of the 1870s was by no means as posh as the drawing room cars.

defects were eventually recognized by the designers of the subway cars, which were built with wide, center doors to encourage speedy and safe entrance and egress and vestibule connections to allow passengers and crew to pass from car to car safely. One reason to explain the elevated's adherence to the old platform car for so long might have come from the basic wood construction of the cars. In the days of wood framing when the elevated roads purchased most of their equipment, a vital structural member was the side truss frames. A center door would have cut this truss in two, requiring heavier floor framing of such massive size that the total car weight would have been too greatly increased.

After Manhattan Railway took over all el operations in 1879, they stayed with the New York Elevated livery: dark-red car bodies, black striping, and the name MANHATTAN in goldleaf on the letterboard. These colors lasted until 1909 when the new operator of the elevated railroads, the Interborough Rapid Transit, decided to simplify car painting by adopting standard railroad Pullman green without any decorative striping.

Then in the early twenties, to make the elevated more attractive, the IRT launched a publicity campaign by advertising their new service the "Open Air Line" and adopting a bright, jazzy color called "goldenrod" for all cars and stations. (Traditionalists, most of whom dispised the bright color, called it "traction orange"!) The idea of capitalizing on the elevated's airy, scenic route was excellent strategy, but it was not well carried out. The shops never quite got around to painting all the cars the new color, so there was apt to be a rather motley assortment of colors on any train. And even those which had been repainted were likely to be rat-colored, because the cars were seldom washed. In 1929 a further repainting program began, changing the cars to a rather drab olive green. By 1933 all the cars had been so repainted except those put in storage due to the drop in traffic brought on by the depression.

The elevated went through the same transitions as steam railways in their lighting of cars. Originally all cars were lighted by sperm-oil candles in brackets on the walls. Then about 1878, when petroleum was becoming available commercially as a cheap and efficient form of lighting, oil lamps were used both in the cars and the stations, though there was some prejudice against them. In case of a wreck or derailment the oil lamps would smash and spread burning kerosene over the highly combustible wooden car and its passengers. Fortunately, there is no record of such a disaster on the el. A much superior lighting system called Pintsch gas was introduced to the United States from Germany in 1883, and the el was one of the first railroads to adopt it. The patented gas was manufactured at the railroad yards and transferred daily to tanks underneath the cars. With the advent of electrification of the line, the trains were easily equipped with electric lights, the naked bulbs in the ceilings being almost enough illumination to let passengers read their newspapers on the way home from work. Until then the practicality of electric lighting was no more real for the elevated than for any other railroad, though many people feel that since it was an urban road electricity was readily available. Even though Edison had been stringing his wires to private customers in the city since 1882, it was still much too expensive, experimental, and undependable for the elevated lines until the entire system was electrified around 1902. Though coal-burning stoves were at first used to warm the cars, the el was a pioneer in adopting steam heating; after electrification they went to electric heat with coils under the seats.

The elevated roads were also early to adopt the Eames vacuum brake, which came out about the same time as the Westinghouse air brake. It was not immediately clear which of the two systems was the better, but the Eames was definitely cheaper. A steam-activated vacuum brake, the Eames system was very efficient for short trains and thus ideally suited for the elevated. And even though the Westinghouse patent overwhelmed it quickly on mainline railroads, the elevated, seeing no good reason to abandon it, stayed with Eames until the electrification. It was a curiosity for North American railways because it had become obsolete years before the el abandoned it, at least as far as regular roads were concerned. After electrification cars were fitted with the Westinghouse air brake.

After their flirtation with the extra-fare drawing-room cars the elevated never bothered to offer anything in the way of unusual cars except for a few rather short-lived bicycle trains and some open summer cars.

The bicycle craze hit New York very hard in the Golden Nineties. Women as well as men found it a very appealing sport, and even society folk took it up, adding further to the vogue. Soon the fad seemed to conquer the entire population. To accommodate the crowds of people who wanted to escape the rough granite cobblestones of the city for the smooth cycling paths of Central Park, Staten Island, and the Bronx, the elevated inaugurated special bicycle trains in 1897, which ran each Sunday up and down Ninth Avenue from 155th Street to Rector Street (since the platforms at South Ferry were too jammed to handle them). The special trains ran Sundays all day except when churchgoers were going to and from services, because the company feared the sight of the sporty heathen bikes might offend the pious. In those days churchgoers were an important factor in transportation schedules. Cyclists carried their vehicles up the steep steps to the station and right into the cars, from which a row of side seats had been removed and bicycle racks installed. Fare was fifteen cents for a rider and his bike or twenty-five cents for a couple and their bicycle-built-for-two. The cars were enormously popular that summer, but the unique service was discontinued at the end of the year and never resumed.

The open summer cars that ran over the elevated beginning in 1902 were an exotic European touch on an otherwise busy, utilitarian commuter road. With their airy park-style benches, gay striped awnings, bright brass trim, and little sliding Dutch doors, the thirty-six summer cars were extremely popular with New Yorkers, who rode them to escape the sultry heat of midtown for cooler Battery Park and the excursion boats at the tip of the island or the pleasant trees and grassy slopes of Bronx Park far uptown. So much sought after were the open cars that they began to cause tie-ups on the line. When a train arrived with such a car hooked on, there was much dashing up and down the station platform by patrons trying to grab a seat. According to New York elevated authority Al Seibel, whose parents rode the el in the days of the summer cars, they were so popular that it was almost impossible to get on one. It is a pity that they were not an operating success because they so clearly contributed very much to the enjoyment of riding the elevated. With America's involvement in the hostilities in Europe, the company had an excuse to rid themselves of the troublesome cars and most were sold in 1918 to Norfolk, Virginia, where they were used to carry shipyard

Anticipating lucrative orders, the Jackson & Sharp car builders sent this fancy parlor car to the New York Elevated as a demonstrator in 1877. Typical of elevated cars of the expansion years, the interior was lavishly paneled and carpeted. COURTESY DIVISION OF HISTORICAL AND CULTURAL AFFAIRS, STATE OF DELAWARE.

tiquated and nasty. However, the Victorian atmosphere of the old coaches, mellowed with an air of wistful decay, was a pleasant tonic for those fevered by the frantic pace on the streets and the depressing, sunless tunnels of the subway—there never was a view of the city like that from an el car. From the vantage point of a window one could survey the desperate slums of Harlem, Ninth Avenue, and the East Side; middle class Tudor City, Chinatown, and the Bowery; the German and Bohemian quarters of Yorkville; the Wall Street financial district; the flat suburban reaches of Brooklyn; the hilly jumble of the Bronx; the quiet, treeshaded streets of Queens; dingy flophouses, dramatic family groups and scenes of great beauty—skyscrapers at dusk, glittering rivers, and dwindling streets.

When demolition of the el began in earnest in 1938 with the closing of the old Greenwich Street-Ninth Avenue line, hundreds of old wood coaches were either scrapped or sold off cheaply. During the war years they went in large numbers for emergency transportation to United States bases and arsenals, ordnance plants, and shipyards. One batch was purchased by the New York City Department of Sanitation and sent up to their employees' summer camp "Sanita" near Peekskill, where they were converted into bungalows. The cars are still on the site, only now used as lodges for a Boy Scout troup. After the Second World War was over, many veteran coaches were dispersed like DP's all over the land, ending their lives, truckless and ignominious, as chicken coops, shacks, and hot-dog stands.

At least four elevated cars from the steam years have survived. Two of the old coaches sent to a shipyard railway in San Francisco during the war have been saved by a local railway historical group. And two old wooden el cars are still running on a tourist railway in the Pennsylvania mountains near Blairsville. The Penn View Mountain Railroad owns two unusual hybrid coaches originally built for Manhattan Elevated by Pullman in 1880 and 1881. They have el car bodies, but the platforms and trucks are from a Pennsylvania Railroad caboose.

workers to the docks. The last three open cars remained in storage until 1938, when they were finally junked.

After public ownership and usurpation by the newer and "more modern" subways in New York, the former bright livery of the remaining elevated cars began to tarnish. In their decline the old wooden steam coaches, some in use for sixty years and long converted to electric traction, continued to clatter up and down the four lines through Manhattan until the very end, taking on a sort of shabby yet genteel decadence that repelled many people who thought the cars terribly an-

10 That Cardiac Climb: Stations along the Line

"No man mad and drunk with liquor can well find his way into the rapid transit cars to insult and terrify women and maim and murder men. A very drunk man cannot climb the stairs." That was an optimistic Victorian sentiment expressed by a New York *Herald* reporter covering the opening of the Sixth Avenue el. He had apparently been taught always to point out at least one moral lesson in a news story, such as the inherent safe haven for New York maidenhood on the platforms of the elevated railroad stations. While there were undoubtedly those who delighted in the scenic high perch of the platforms, the cardiac climb up three flights of slippery iron stairs was not for most travelers the happiest part of riding the el trains.

Despite their high stance the elevated stations received almost universal praise from the press over the years because they were thought to wed so perfectly both utility and beauty. When the

fancy second-generation stations were bright and new in the 1880s, they were considered to be models of contemporary architecture, enchancing the cityscape with their tastefully decorated gables and graceful iron stairways. The stations built during the el's expansion years from 1878 to 1882 were vast improvements over the primitive wooden sheds of the cable generation and first years of steam operations, being larger, more comfortable, and more visually attractive than before. Although the station buildings of the Metropolitan Railroad on Sixth and Second avenues and the New York Elevated on Ninth and Third avenues varied somewhat in plan and detail, they looked essentially alike. (In fairness the stations along Sixth Avenue were probably a bit grander than the others.) After Manhattan Railway introduced a common color scheme and standard materials and plans for repairs, exten-

The spreading eaves and staircases of the elevated stations created a safe haven for sidewalk vendors of all sorts. The decorative appearance of the stations did much to eclipse the dark intrusion of the el in the streets. COURTESY PRINTS AND PHOTOS DIVISION, LIBRARY OF CONGRESS.

sions, and modifications on all el lines in the mid-eighties, the stations tended to lose most of their former corporate identity, taking on a single elevated-railroad character. They were all painted a fresh apple-green outside with dark, hunter green and maroon trim, a color scheme that was continued by Manhattan and Interborough Rapid Transit (IRT) until the Open Air Line orange paint job came along in the early 1920s.

In many respects the el stations were hardly different from many mid-nineteenth-century railroad depots, but in the process of being hoisted up in the air a transformation took place that made them much more prominent, interesting, and pic-

turesque. Quintessentially Victorian, the general design of the station exteriors might be classed as Hudson River Gothic, and there is a close resemblance to the romantic cottages of Alexander Jackson Downing. The best-known personality associated with the el stations was Jaspar Cropsey, architect of the stations along the Sixth Avenue line. A fashionable landscape painter of the time, Cropsey was a curious choice for the assignment, since he was neither a civil engineer nor experienced in the new technology of the elevated railway. Apparently the directors of the Metropolitan Road recognized that his name had a certain cachet and considered him the perfect man to furnish the elegance and artistic style they wanted for the stations. Cropsey had studied landscape painting and architecture for several years in New York before moving to London for further studies. There he was a regular exhibitor at the Royal Academy and well acquainted with the coterie of Gothic enthusiasts led by John Ruskin. After Cropsey returned to New York, his highly romantic scenic paintings drew attention from the city's Fifth Avenue and Gramercy Park set, and he accepted occasional commissions to design houses, though his chief interest remained painting. A possible connection with his commission to design the elevated stations might have been his association with George Pullman, who not only backed the road financially but built many of the cars. Cropsey had also supervised the construction of Pullman's houses in Chicago and Long Branch.

Jaspar Cropsey's reputation remains well intact today, his paintings hanging in such prominent collections as the White House, the Metropolitan Museum in New York, the Corcoran Gallery in Washington, the Wadsworth Atheneum in Hartford, the Boston Museum of Fine Arts, and the Toledo Museum of Art. In 1970 the National Collection of Fine Arts in Washington held a one-man show celebrating the artist's career. The catalog of the show is still in print and well worth reading. In addition to his prowess as a landscape artist Cropsey deserves high praise for working out the solution of easy access to the new mode of transportation, the elevated railroad. His covered pavilion staircases rising in graceful tiers up to the waiting rooms are superlatively decorative, yet practical as well, and his contribution to New York street architecture did much to eclipse the dark instrustion of the el in the streets.

From the pedestrian walkways the el stations looked incredibly rich and inviting in their pretty, ornamental gingerbread, romantic peaked gables, lacy iron balustrades, fancy eaves, quaint cupolas topped with finials, and Corinthian capitals on iron

A Third Avenue elevated station, 1948. In cold weather a traveler could still pause and warm himself by a glowing potbellied coal stove and savor for a moment or two the nostalgia of old times past. PHOTOGRAPH BY ARNOLD EAGLE, COURTESY THE MUSEUM OF THE CITY OF NEW YORK.

In many respects the el stations were hardly different from many mid-nineteenth-century railroad depots, but in the process of being hoisted up in the air a transformation took place that made them more prominent and picturesque. Eighteenth Street and Third Avenue, 1942. COURTESY PRINTS AND PHOTOS DIVISION, LIBRARY OF CONGRESS.

The blue stained-glass windows in the stations remained for many years as reminders of another era. PHOTOGRAPH BY ARNOLD EAGLE, 1948, COURTESY THE MUSEUM OF THE CITY OF NEW YORK.

pillars. Unlike most rail depots of the age, the new stations were built almost entirely of iron, a requirement of the Rapid Transit Commission, presumably as an effort toward fireproofing. Yet the buildings were so light and delicate architecturally that few people would fail to think of them as rustic wooden cottages.

Passengers reached the station platforms from street level via light, iron stairways enclosed on the sides and covered by fancy pavilion roofs. The stairs were particularly graceful, some with elaborate ironwork and sculptural decorations in a lion's head motif. At the top of the stairs was a balcony from which passengers entered the ticket office and waiting rooms, one each for men and women. The waiting rooms, according to *Leslies Illustrated,* were done in the Eastlake style with black walnut paneling and pine benches stained and grained in various hues. The term *Eastlake,* which has been used to describe the elevated's cars, dummy locomotive cabs, and passenger stations, is derived from Charles Eastlake, a

The platforms at Thirtieth Street were typical of many Ninth Avenue el stations. Photograph taken in 1906 shortly after electrification. COURTESY THE SMITHSONIAN INSTITUTION.

After buying a ticket, a passenger could drop his ticket into the box in the charge of the attendant and go out to the covered platform and watch the bustling life of the city in the streets below. COURTESY THE COLLECTION OF ROBERT M. VOGEL.

Station for the City Hall branch on Park Row shortly before the Brooklyn Bridge terminal (left) was completed. The large building at the right is the Staats-Zeitung, New York's German language newspaper. At the far left is the Bogardus shot tower, torn down in 1908 during subway construction.

When bright and new in the 1880s the gingerbread gothic el stations were considered models of architecture, enhancing the cityscape with their tastefully decorated gables and peaked pavilion roofs. This is the Forty-second Street station on the Sixth Avenue line. COURTESY Harpers Weekly, JULY 20, 1878.

fashionable English authority on the decorative arts in the 1870s and 1880s. His book *Hints on Household Taste* was a best-seller in this country as well as in England. Eastlake was highly critical of the classical styles of Greece and Rome that had been so long in vogue and urged a return to what might be termed Victorian Gothic. The American version of Eastlake design usually refers to picturesque gingerbread cottages and heavy walnut or mahogany furniture with lots of geometrical carving based on the carpentry of the Middle Ages.

After buying a ticket at the station agent's office, a passenger could if he liked drop his ticket in the box in charge of the attendant and go out to the platform since it was well covered and made a

pleasant promenade for watching the bustling life in the streets and avenues of the city below. But it was usually just a few minutes before a train would come along, and since the station platforms were flush with those of the cars there was no climbing up and down awkward steps in getting into a train.

Normally the pattern was to build two separate stations at each stop at intervals of about five blocks downtown and about half a mile uptown where the population was somewhat less dense. The up stations were on the east side of the street and the down on the west so that passengers would not have to cross the tracks to get to their trains. Station buildings were placed over the street

The spidery stairs lifted gracefully up to the station at Ninth Street and Third Avenue, 1882. Though built entirely of iron, the covered stairs had an air of lightness because of the lacy balustrades, ornamental scrolls, and slender columns.

crossings with the stairways leading down in pairs to the cross streets. An alternative was to build a single island station between the tracks such as in the lower East Side, where there was not room for two separate structures. The singe island stations had the advantage of requiring fewer employees than the double ones, though unless special platforms were constructed there was apt to be great confusion to both incoming and outgoing passengers. Also, the island stations required a wider track right-of-way along the line and interfered with the possibility of using separate express tracks.

One of the most evocative descriptions of a New York elevated railroad station comes not from a contemporary journal but from a remarkable modern novel set in nineteenth-century New York. Jack Finney's *Time and Again* captures the romance of riding the el the way few other books have. Writing in the first person he says; "As I climbed the steps, even the ironwork of the railings seemed wonderfully familiar. I'd visited New York often as a boy, ridden the El many times. And now here again, inside the little station, were the bare worn floorboards, the wooden tongue-and -groove walls, the little scooped-out wooden shelf projecting from under the change-booth window, grained and polished from ten thousand hands. There was a cuspidor on this floor and the station was lighted by a single tin-shaded kerosene ceiling lamp. I shoved two nickles in through the little half-moon hole at the bottom of the wide-meshed grill between me and the mustached man in the booth. He took them without looking up from his paper, and shoved out two printed tickets. Then we walked out to the platform, . . . to see the dozen or so passengers: the women in skirts that nearly brushed the platform,

El stations were built over street intersections with the stairways leading down to the cross streets. Early stations, such as this one on Third Avenue at Twenty-third Street, were built with only one stairway. The second was added later and did not always quite match the original one. Ca. 1882.

and grew dirty and unkempt. In time commercial posters found their way up to the platforms, advertising everything from the latest gypsy band to cigars, baking powder, and cornflakes. Coin-operated vending machines dispensing peanuts, candy, chewing gum, and snuff were installed in the late 1880s, and news vendors and shoeshine stands sprang up everywhere on the streets along the el's route, taking advantage of the cover offered by the cantilevered staircases. With the vastly increasing number of passengers about the turn of the century the company installed escalators in 1901 at some of the busier midtown stations in the shopping district: near Macy's Department Store at Herald Square and another adjacent to Bloomingdale's at 59th Street on the Third Avenue el. And at the high section at 116th Street and Eighth Avenue an electric elevator was put in to whisk patrons up to the track. Such new and wonderful mechanical conveniences added greatly to the vibrance and thrill of elevated railroad travel.

As they endured for many years into the twentieth century, the picturesque little depots grew rusty—even a bit grimy—but few people with any eye for architecture could resist the charm of those quaint cottages. Their blue stained glass, rich, gingerbread fretwork, and medieval pavilion roofs tended to soften the rawness and stridency of urban street life through depression, world war, and prosperity. After as long as seventy-five years a handful of el stations remained as wonderfully quaint anachronisms lying in modern steel and glass canyons, where in cold weather a traveler could pause inside to warm himself by a glowing, potbellied coal stove and savor for a few moments the heady nostalgia of times past.

wearing bonnets and shawls, some carrying muffs; and the whiskered men in their derbies, silk hats, and fur hats, smoking cigars, carrying canes.''

With the heavy traffic of thousands of feet rushing through them every day the pristine beauty of the levitated gothic cottages eventually diminished. Though rather pretty in themselves, the stations began to show the effect of overuse

11 Life on and along the El

No trip to New York was complete without a ride on the world-famous elevated railroad. There was no pleasanter way of seeing the city than from that high route. Guidebooks advertised the wonders of the el—many with illustrations and lavish descriptions—pointing out the grand panoramas and vistas to be had of the populous polyglot city, all for the price of a five-cent ticket. Once on board the stranger from Sioux City or Nashville found himself in a handsome, airy, and comfortable car, embarked on a thrilling journey he would probably always remember, whirling along the streets on a level with the second stories of houses—if not higher—looking down onto incredible sights in the teeming pavement and sidewalks below. Beneath him horsecars passed with an echoing jingle of harness bells, and in the distance were sparkling river views crowded with shipping, boisterous Bowery barrooms, rows of dignified, brownstone town houses, tall church spires, and green parks. It was something to write home about, like a visit to the Statue of Liberty, Barnum's Museum, or later the Empire State Building.

The following trip on the Sixth and Third Avenue elevated lines has been gleaned from several guidebooks to New York City, primarily of the late 1880s and 1890s, though there are some incursions into later decades. The idea, as with the original sources, is to capture the experience of riding on the el and to show the neighborhoods and points of interest traversed by the trains. Since the ride does not represent any single point in time, there is apt to be a historical clustering, anywhere from 1882 to 1930; however, the emphasis is primarily on the late nineteenth century.

We shall begin our ride on the Sixth Avenue elevated, the main West Side line, at South Ferry station, southern terminus of all the elevated roads

From shadowed intimacy a four-car el train emerges into sunlight on the lower East Side. The contrasts were striking, the sense of historical continuity rich.

and a busy ferry junction. Leaving the big turreted ferry house, which was for years a showplace in the city, our train turns left through the treetops of Battery Park near Castle Garden immigrant reception center, to Battery Place station, near the Cyrus Field Building, the Produce Exchange, many foreign consulates, and steamship offices. The pace of our train is brisk as we dart up New Church Street—then in only a few moments we stop quickly at Rector Street Station, named for the first rector of Trinity Church. Because this is the station for busy Wall Street financial offices, a rush of passengers stand waiting on the platform just outside our window to board our car. While our caravan pauses to take on new patrons, a scarf of purple smoke rolls back from the locomotive and in a downdraft clouds our view. But as the dark cloud vanishes, we look up and see old Trinity Church with its churchyard, burial ground for many prominent New York families. Next the burly, uniformed conductor barks out, "All out for Cortland Street," and we grind into another bustling depot in commercial downtown Manhattan, the station for swarming Washington Market, where thousands of wagons and drays stand in a disorderly array to sell their wares. Nearby are the Victorian chateau called the Coal & Iron Exchange, Maiden Lane, and scores of brick and brownstone commercial blocks. Immediately after pulling out from the station, we pass under the graceful Georgian spire of St. Paul's, oldest church building in the city, and just beyond the massive temple of St. Peter's.

We hear the low, whooshing sound of the vacuum brake under us, for Park Place station is now coming up—stop for Newspaper Square, the post office, City Hall, and the court house. Here our track turns sharply one block west through Murry Street (it is said to be the sharpest railway curve in the world), and our wheels screech painfully from the tortuous bend before coming to a halt at Chambers Street station, where several passengers alight for the Fall River and Providence steamboat lines and the Erie railroad ferry on the Hudson River slips nearby. The course of our train is now up West Broadway through the wholesale grocery and dry-goods district—an area crowded with massive warehouses, hotels, and stores. Franklin, Grand, and Bleeker Streets are the next stops as we near elegant Washington Square with its dignified Greek Revival mansions, which seem curiously diminutive from our odd, second-story perspective. Turning west again through Amity Street, our locomotive passes another el train coming from uptown, and the engineers sound their tenor steam whistles in salute. We then reach the foot of Sixth Avenue, principal right-of-way of this elevated line, stopping at Eighth Street in the heart of Greenwich Village, an artsy Italian-Bohemian quarter, the Soho of New York. This station rests under the shadow of Jefferson Market Courthouse, cast by its tall Gothic clocktower. Built in 1876 under the direction of architects Vaux and Withers, the picturesque structure serves as market, police court, and prison. (The building is still standing today, having been remodeled in 1967 as a public library and retaining all of its neo-Gothic details both inside and out.)

At the next station, Fourteenth Street, many shoppers leave the train for Macy's Department Store and nearby Union Square. Along this stretch of Sixth Avenue between Fourteenth Street and Twenty-eighth Street begins New York's principal shopping district of the late 1880s. As far as Fifty-ninth Street where it is halted by Central Park, the broad avenue is solidly built up with handsome stores and shops, theaters, restaurants, and hotels. On its upper blocks are stylish flats rivaling fashionable Broadway. The sidewalks here are constantly filled with throngs of smartly dressed shoppers attracted by fine displays of goods and the belief that Sixth Avenue prices are lower than on Broadway. All through the day the

At the upper end of the Bowery stands the tall brownstone Cooper Union and the busy Ninth Street elevated station. C. 1900. COURTESY THE COLLECTION OF E. A. SEIBEL.

avenue is bright and lively with people, and the passage of the elevated trains on the railway overhead adds greatly to the gaiety and spectacle. At night the many colored lights of the el stations lend another attractive feature to the scene and the whirl and roar of the brilliantly illuminated trains as they whiz by overhead give to the street an air of life and bustle in keeping with the movements of the crowd on the sidewalk below.

Near the Eighteenth Street station stands Chickering Hall, where in 1882 Oscar Wilde delivered his first American lecture, his massive form clad in full dress coat, white vest, and black knee breeches and silk stockings—everyone in New York society was there. At the Twenty-third Street el depot

passengers alight for Ehrich's Department Store, the Grand Opera House, the great, gray Masonic Temple, and famous Booth's Theater, where Mme. Sarah Bernhardt, the greatest tragic actress of the age, made her American debut in *Adrienne Lecouvreur.* With the inexorable movement of the city uptown, the commercial hub of New York gravitated up Sixth Avenue, so that by the turn of the century or thereabouts the elevated station at Twenty-third Street was one of the busiest and most elaborate of the entire system.

Broadway crosses Sixth Avenue at Thirty-fourth Street, creating at the intersection a large, open space of two triangular parks called Herald Square because of the prominent New York *Herald* building. The lower park contains a statue of journalist Horace Greely, founder of the *Tribune,* and the upper wedge of the statue of

In the Bowery the elevated superstructure created a sort of latticed arcade over the sidewalks, which seethed with domestic and commercial life. A derbied roast-potato vendor plies his trade near Chatham Square. COURTESY THE MUSEUM OF THE CITY OF NEW YORK.

William E. Dodge, a merchant, philanthropist, and early supporter of elevated railways for the city. It is a nice touch to think of the el trains rumbling by the statue of the visionary man who was so early to recognize their rich potential. (His statue now stands in Bryant Park.) Herald Square is a very congested area, coming to prominence at the turn of the century as one of the premier retail districts in the nation. In 1902 Macy's opened their colossal new emporium here, closely followed by Saks and Gimbel's. The Pennsylvania Railroad's mammoth new station was built in 1910 just a block or so west, connecting Manhattan with New Jersey and Long Island via deep, assertive tunnels under two rivers. In the 1890s the area around Greeley Square was called the Hackman's Poorhouse because of the large cabstand there. Polite coachmen in canary-yellow topcoats and shiny silk hats with spick-and-span hansom cabs park there by day, and at night it is inhabited by night hawks in shabby old coats and slouch hats ready to take clients to Tenderloin dives and brothels.

Elevated trains flash close by Herald Square's main attraction, the New York *Herald* Building, designed for publisher James Gordon Bennett by the celebrated firm of McKim, Mead, and White. The palazzo is one of the city's architectural wonders, with its Venetian arcade and the row of bronze owls around the cornice that wink their electric eyes at night. On the roof above the main entrance is the famous clock with the bronze figures nicknamed Stuff and Guff, who strike the hours. Great plate-glass windows set in the arcade on the side of the building are usually crowded at night with people watching the big presses at work throbbing out the morning edition.

Continuing our ride after stopping at busy Herald Square, we pass solid blocks of stores, approaching Bryant Park between Fortieth and Forty-second streets, a large, green square usually full of playing children during the day. Behind it used to sit the grim Egyptian revival walls of the Croton Reservoir, principal holding source of New York water. (Now the site is occupied by the

Sixth Avenue from Fourteenth Street to Twenty-third was one of New York's premier shopping districts in the 1890s. All through the day and night the avenue was bright and lively with people. And the elevated trains added to the gaiety and glamour. A holiday spirit pervades this old stereoptican photograph. Flags wave from the windows of B. Atlman (left) and Siegal, Cooper, and Company (right).

On the West Side at Ninety-ninth Street Manhattan Island is rather hilly, and the el structure quite high. Trains rumbled along on a level with the fifth-story windows of nearby apartment buildings. Because of the great height the trestle was lighter, airier, and less obtrusive than farther downtown. Date of the photograph is February 27, 1891. COURTESY THE NEW YORK HISTORICAL SOCIETY.

New York Public Library.) At the Forty-second Street el station passengers alight for Grand Central Depot two blocks east, the Metropolitan Opera House, and Broadway theaters a block to the east. Many gentlemen's clubs, Temple Emanu-el, the Church of the Heavenly Rest, and Holy Trinity Church on Fifth Avenue are within easy walking distance. The Fiftieth Street station is the stop for many sights on upper Fifth Avenue: the Windsor and Buckingham Hotels, Columbia University, St. Lukes Hospital, fashionable St. Thomas Episcopal Church, the great Gothic edifice of St. Patrick's Roman Catholic Cathedral, and the ostentatious town houses of the Vanderbilt

family. (In the 1930s the old Victorian elevated station was adjacent to the sleek, modern Radio City Music Hall, part of the Rockefeller Center complex—a wonderful juxtaposition.) Here passengers change for Central Park, unless they are on a Central Park train, which will carry them straight ahead to Fifty-eighth Street and the park entrance.

In a noisy clattering over switches, our Harlem-bound train turns westward through Fifty-third Street to Ninth Avenue, then runs northward up that avenue parallel with the Ninth Avenue elevated line. Traveling up the West Side on Ninth Avenue after recrossing diagonal Broadway, we find the region primarily residential and peaceful compared with hectic Herald Square. To the right between tall apartment houses we catch glimpses of Central Park, whose hills and trees stretch like an undulating green belt. There the prestigious gable-roofed Dakota flats rise on

Eighth Avenue, so called because when it was built people said it was so far from civilized New York as to be in the Dakotas. This quarter along Central Park grew quite fashionable after construction of the elevated railroads, making the area much more accessible than in the days of horsepower. Before 1880 and even afterwards vacant lots, open fields, and farms were located here, occupied by squatters whose ramshackle homes, goats, and multitudinous children added what one might now consider a picturesque touch to the scene, but which at the time was thought to be a blot on the landscape. Since 1890 costly residences and elegant apartment buildings have been built, inhabited by many distinguished citizens, including Civil War General Sherman. Just before reaching the Eighty-first Street el station we pass the castlelike American Museum of Natural History, repository for vast worldly collections of flora and fauna.

Around the Ninety-third and 104th Street stations the island is rather hilly, and our car rumbles along sixty-three feet high over the pavement. The danger only adds to the thrill of the ride. At 110th Street the el track makes a double turn known as "Suicide Curve," nearly touching a corner of Central Park, and heads on up Eighth Avenue on the high iron trestle that De Lesseps, engineer of the Suez Canal, called "an extraordinarily audacious bit of civil engineering." Because the ground is low here, we are carried across on a level with the fifth-story windows of nearby apartments in

On the East Side the el tracks ran through the hub of old New York, offering passengers waterfront views of masted ships on the East River. Many of Manhattan's oldest buildings survived in this quarter, such as in this reminder of the city's Dutch ancestry. The scene is looking down the ancient Barger Path and Old Slip, 1902.

Elevated trains flashed through the heart of commercial New York at Herald Square. In the center is the Herald Building. Close by were such emporiums as Saks, Macy's, and Gimbel's. The uptown station in the foreground sat right inside Greely Park.

whose windows we see the faces of women and children watching us watch them. They do not seem to mind our intrusion into their homes but reflect only a curiosity at our traverse.

The tracks are so high at the 116th Street station that passengers must take an elevator to reach the platform. The ride here is rather scary but the view sublime: Harlem to the east and north and the upper end of Central Park on the right, which appears as a green grove. On the left are the ornamental stairways of Morningside Park and the high pillared front of the Leake and Watts Orphan Asylum, on whose site the Anglican Cathedral of St. John the Divine was afterward erected. In the distance are the trees and roofs of the great Bloomingdale Asylum (later to be the new site of Columbia University) and the mansions on the heights overlooking the Hudson River.

South Ferry was the southern terminus for all el lines, trolleys, omnibuses, and Manhattan landing for mainland Long Island and Staten Island ferries. The Coney Island Ferry (left) took passengers directly to Buffalo Bill's Wild West Show. At the right are the tip of Battery Park and the turreted Barge Office Building. Governors Island is in the background. Photo 1894. COURTESY PRINTS AND PHOTOS DIVISION, LIBRARY OF CONGRESS.

A few blocks beyond our train crosses St. Nicholas Avenue, the ancient country road to King's Bridge and Albany, stopping briefly at the 125th Street station, where rocky ridges, market gardens, and remnants of shantytowns diversify the dead level and uniformity of the graded streets of lower Manhattan. At the left the ground rises into a long ridge called Washington Heights, and near the 135th Street station is the Roman Catholic convent of the Sacred Heart. The engine houses and shops of the elevated railroad company are at the 145th Street station, and half a mile beyond it the terminus of the line with the carriage bridge on the site of the old Macomb's Dam. In this neighborhood are a number of large picnic grounds, dancing halls, and beer gardens, which in the summer are very popular with the citizenry. Nearby stairs lead up to the level of Washington Heights and to the Edgecombe Road that goes to High Bridge reservoir and park and the lofty arches of the aqueduct. This last station is also terminus of the Northern New York Railroad, which would take us, if we chose, on northward to the villages of Westchester County.

* * *

On the East Side of the city the main elevated road is the Third Avenue line. The East Side from the earliest time was the cradle of mercantile life. The Dutch founders of the city settled here, locating their canal on Broad Street and anchoring their vessels in the East River, on whose banks their primitive wharves and storehouses were built. There is not a street between the Battery and City Hall Park that is not redolent with the romance of the old merchants of the metropolis. After boarding our train at South Ferry, we enter the heart of historic New York along the East River, a route teeming with immigrant tenements, sailing ships, spice and coffee merchants, importers, and chandlers engaged in the busy and profitable trade that has made the city rich and great. The train runs along ancient narrow Water Street, Coenties Slip, Front Street, and Old Slip, offering dramatic views of the busy river and across to the Brooklyn waterfront. For the el traveler there are many architectural surprises, because much of Manhattan's earliest building was concentrated in this section, having survived for more than a century. From the sunlit intimacy

of small-scale Georgian buildings our train turns into the shadows of new tall drug and tobacco warehouses in Pearl Street. The contrasts are striking, the sense of historical continuity rich and rewarding. One is only a block away from the colonial era, for at Broad Street we pass Fraunces Tavern where General Washington took leave of his officers in 1783.

Winding through the gorge of Pearl Street, we pass Fulton Street close to the enormous fish market and the ferry to Brooklyn, the principal route to that city before the Brooklyn Bridge was opened in 1883. The street here is so narrow that the Fulton Street el station had to be built into the old United States Hotel in order to secure a platform. Next we approach Franklin Square almost under the approach to the Great Bridge; this is the stop for the steamboats to New Haven, Bridgeport, and eastern Long Island. At the end of the eighteenth century this square was a very fashionable quarter surrounded by the mansions and gardens of the wealthy and well-connected Walter Franklin family. In a house on this square General Washington and his wife lived for a time, holding official receptions after his inauguration as president. At that time lower Pearl Street, through which our train has just passed, held the homes of many aristocratic families who before the Revolution led New York society: the De Lanceys, Livingstons, Morrises, Bayards, De Peysters, and Congers. Suddenly our train skirts closely by the imposing iron facade of the publishing house of Harper and Brothers; this was the most elaborate commission of architect James Bogardus, who invented the cast-iron building and foreshadowed the use of steel in modern building construction. Just to the west is the leather sellers' district, known as the swamp ever since the days when it was a brush-hidden morass on William Beekman's farm and was surrounded by the ill-smelling vats and yards of his tanneries.

Under the immense stone arches of the Brooklyn Bridge approaches, just north of the square, lie dark, mysterious vaults used as wine cellars years before the bridge was ever finished. Dark, gloomy, and moldy, the cellars house the choicest wines of the city: light, dry Pommards, Chambertins, and Chateauneuf-du-Papes from France; brisk Reislings and Moselles from Germany; elegant sparkling champagnes. Year after year dust and mold has collected on these bottles as they wait in tiered

Clearing the streets after the Blizzard of '99. The scene is midtown on the Sixth Avenue elevated—probably near Herald Square. Though the winter storm is over, the photographer of this commercial shot has simulated snow by speckling the picture with white paint. COURTESY PRINTS AND PHOTOS DIVISION, LIBRARY OF CONGRESS.

racks to please the palates of the smart Fifth Avenue set. Known as Oech's Wine Cellars, they are a famous sight in New York, attracting hundreds of visitors every year. (With Prohibition in the 1920s the vaults were forced to disgorge their luxurious collection of wines, taking instead a noisome cargo of fish from the Fulton Fish Market. The romantic era was over.)

From Franklin Square up New Bowery our train passes through the rough region of the Fourth Ward. On the right is a remnant of an old Sephardic Jewish cemetery dating from 1742. Here there are many queer old streets now full of Irish tenements, which a century or less ago were the residents of more gentle folk. Along this stretch of the el are marvelous views of John and Washington Roebling's heroic Brooklyn Bridge with its great Gothic piers and graceful cablework tracery. Chatham Square is coming up, busiest elevated station in the city, with a branch for City Hall. It is a crowded, noisy, confusing place for strangers, surrounded by factories and tenement houses filled with hordes of foreign laborers. Exotic Chinatown is but a block or two west.

In a soft chuff-chuffing our locomotive pulls us smoothly away. Once past the station there begins that famous, or shall we say infamous, region called the Bowery. It would take a combination of Dickens's tongue and Hogarth's pen to describe it fully. Originally an Indian trail, the Bowery, which runs from Chatham Square to Cooper Square, was known in Dutch days as the "road to the *bouwerij*

114

[farm]'', Peter Stuyvesant's country estate, which extended to the East River from Sixth to Eighteenth streets. During the Revolution the Bouwerie Lane became Bowery Road, and in 1807 it became the Bowery. The Bowery is an odd, wonderful place, half humorous and half dreaded, strange and foreign to the tourists, a quarter to be visited but never known—a vast and insurmountable barrier to respectability, filled with beer halls, Irish gangs, and riffraff. Some of the dives that dot the street have never been equalled for the frightening and deadly quality of their liquor. In many of the lower-class barrooms drinks are three cents each, and no glasses or mugs are used. Barrels of fiery spirits stand on shelves behind the bar and pour out their contents through lines of slender rubber hoses. The customer having deposited his pennies on the zinc counter takes an end of the hose in his mouth and is entitled to all he can drink without breathing. The moment he stops for a breath the watchful bartender turns off the supply.

Looking out on the Bowery life from our elevated perch, we pass a continuous line of theaters (most not respectable), cheap museums, boardinghouse lodgings, and small shops. The din of traffic and life is tremendous, but the feeling of intimacy is strong. The second-story windows beside us bulge with leaning women and their children, gray pillows, dirty curtains, stained bedding, and men in their underwear reading at kitchen tables. Near Canal Street we pass very close by the pillared front of the old Bowery Theater, quasirespectable in the dress circle but positively swarming with indecent ragamuffins of all degrees and sexes elsewhere. (In 1879 the name was changed to the Thalia Theater for German drama; then in 1892 Hebrew performances dominated, until the next ethnic upheaval fifteen years later, when the Italians took charge.)

As our train stops briefly at Canal Street, the noise below becomes terrific. At night the windows gleam, the saloons are all aglare; tinny pianos and violins send their airs into the night in an instrumental discord that attunes itself outside to the roisterers' songs, the brawlers' oaths, and the hundred strange-tongued voices of the street. Then the gates of our train clank shut, and once again we move off uptown; the wheels drown the noise, and we are left with silent, darkened, second-story windows beside us. Grand Street,

principal shopping street on the East Side, is behind us as we reach Houston Street, the horse-car connection for Police Headquarters, the North River ferries and eastward into the heart of the great East Side tenement and factory district, home of the German, Jewish, Russian, and Bohemian population of wage workers. Since 1855 to 1892 nearly eight million aliens have entered the country through the big immigration station at Castle Garden in lower Manhattan, and it seems that many have chosen to live in this fiercely congested slum on the lower East Side.

Coming up ahead at the upper end of the Bowery is a tall brownstone building called the Cooper Union. Erected in 1854 by Peter Cooper for the improvement and education of the working people of the city, this free school has three thousand young people enrolled, mostly in night classes, since most students work at their trades during the day. It was in the Cooper Union's great lecture hall, all draped in black mourning, that Irish residents of the city in 1887 held their anti-Jubilee in honor of Queen Victoria's fifty years of misrule; meanwhile, the British residents celebrated the Jubilee at elegant Metropolitan Opera House. Ironically, Abram Hewitt, Peter Cooper's son-in-law and mayor of New York, was the principal speaker at the Anglo-American celebration.

In the little triangular park looking down the Bowery from the front of the Cooper Union sits a statue of the venerable philanthropist by sculptor August St. Gaudens, once a student at the institute. The figure looks serenely out upon the busy elevated railroad he had done so much to encourage. As a friend and supporter of Charles Harvey's cable elevated, Cooper had done as much as any man to encourage cheap, efficient travel for the people of New York. Our train passes directly beside the Union, entering Third Avenue by a slight curve and stopping beside the large Ninth Street station to let a covey of passengers off for the Astor Library, Cooper Union, St. Marks-in-the-Bowery Church, and the many stores along nearby Broadway. The most famous of these stores was the fashionable mercantile establishment of A. T. Stewart, which covered the entire city block between Ninth and Tenth streets. Later John Wanamaker of Philadelphia took over the Stewart Store and added another block-square building to the south.

Once past the colorful Bowery the ride up Third

Avenue is less turbulent. Between Fourteenth and Eighteenth Streets we see tall churches, theaters, shops, and small factories. At Thirty-fourth is the el branch for the Long Island Railroad and Manhattan Beach. At Forty-second Street another branch leads free of charge to the mansarded edifice called Grand Central Depot and the Grand Union Hotel, which contains more than five hundred sleeping rooms and is the most popular hotel in the city. At Sixty-seventh Street passengers leave our train for Lenox Hospital, the Normal College, and the Central Park menagerie; at Eighty-fourth Street more leave for the Metropolitan Museum of Art and the Egyptian obelisk (or Cleopatra's Needle), presented to the country by the Khedive of Egypt. The remaining course up the avenue calls for little description, since we look out mostly on a medley of roofs and gables, closed shutters, and draped windows. The street we traverse is the old Boston Road; straight and populous, it is solidly built up with a continuous line of retail shops and lofty apartment buildings inhabited by well-to-do and respectable people, largely of German heritage. The elevated Stations at Eighty-ninth, Ninety-eighth, 106th, and 116th streets are in the old Dutch community of Harlem, whose main thoroughfare we reach at 125th Street, where some passengers of our train transfer to a cablecar and go west to the Sixth Avenue el and on to Washington Heights or to 129th Street and change to the cars of the Suburban Elevated Railway and into the Bronx.

As for us, our eyes and ears are too surfeited to ponder further travel for a while. Instead, we leave the little el train, climb down the pretty iron stairs and meander slowly on the pavement in search of a pot of hot tea. Our ride on the el is finished, but the images remain sharply engraved in our memory. It is a remarkable adventure, filled with incredible human drama.

12 That Uproar to the Eye

How the elevated railroads could ever be built, let alone endure for as long as they did, is astonishing, for the public and private outrage against racing dirty, noisy locomotives down the middle of four of the principal avenues of the city was enormous. The forces of opposition had been very great even against the relatively clean, relatively silent, and relatively short cable elevated on Greenwich Street, but when plans were announced for the completion of a citywide el network New Yorkers saw the coming of a calamity so dire as to destroy the very life of their city. Whereas objections to the infant cable elevated were made by occasional property owners, horsecar companies, and a group of downtown businessmen led by A. T. Stewart, who were interested in keeping select streets such as Broadway and Fifth Avenue free from mass transit, the opponents of the citywide system were legion. The press stirred up public

sentiment against the railroads; mass meetings were held to turn out the populace; lobbyists were hired; and injunctions were sought to halt this dreadful encroachment to municipal privacy. Property owners complained bitterly that the locomotives belched nasty, black smoke into their windows, soiled laundry hanging on their clotheslines, shed hot ashes on their heads, made horses stampede, cut off sunlight, and destroyed the value of their homes and stores.

Opposition to the el was quite well organized. The clique arranged large public meetings at Chickering Hall to exhort the crowds and attract new recruits to their crusade. At one such meeting it was resolved that "we utterly repudiate the idea that the property of one man is to be sacraficed to put money into the pockets of another . . . and we believe that the so-called elevated railroads are a sham and a snare" to our city. Speaking before the

The elevated was an unsightly blemish to many old neighborhoods, blocking some magnificent wide avenues with its great columns and creating interminable dark tunnels where there had once been sunlight.

crowd, Mr. Egbert said that New York was on the verge of an impending disaster greater than it had suffered in three hundred years, greater than any past pestilence, "be it cholera, yellow fever, conflagration or war . . . greater than all of those combined, because it is permanent injury." City Assemblyman Isaac Hayes struck a sentimental note by recalling the cheery tinkle of the horsecar bell and the sturdy plodding of the faithful horses through fair weather and foul. Another witness declared that the elevated structures would be extensively patronized by suicides, and thus dangerous to pedestrians. Chief attorney and lobbyist for the property owners, William Evarts (soon to take up the post of United States Secretary of State) appealed to the city fathers to adopt the subway plan instead of the elevated. The

opponents circulated pamphlets around neighborhoods to be affected by the roads, illustrating the horrors of such travel. One old print shows horses being frightened by the iron monsters and stampeding into the posts and trampling children and adult pedestrians while steam from the engine sears the mangled corpses.

One tends to sympathize with the property owners and renters, for living along the route could not have been pleasant. Though there may in retrospect be an abstract beauty to the diminishing perspective of the structure with its lacy bower of iron supports receding far into the distance, the reality was no doubt far less romantic. The elevated was an unsightly blemish to many neighborhoods, blocking some magnificent, wide avenues with its great columns and creating interminable, dark tunnels where there had been sunlight. It darkened the streets and lower stories of many stores and home, particularly at the corners where stairs and stations were built, and the shade kept the streets damp and wet long after others were dry. The *New York Journal of Commerce* called the elevated a "gross injustice," and the *Scientific Americam,* outspoken critic of the roads for years, labled them "an oppressive nuisance." To the New York *Times* the tracks created a "perpetual city of night."

The tracks and columns of the elevated absolutely spoiled the beauty of many fine buildings—insolently drawing its thick, black line across the Corinthian fronts of old theaters, grazing classical pediments, flouting elegant fenestration, and debauching handsome churches. Characteristically, Russell Sage lightly dismissed all talk of aesthetics by stating categorically that the primary use of the streets of the city was for transit and not for the display of architectural features of the buildings lining them. Another elevated railway backer said the tracks were actually a godsend, because they hid so much from sight and kept one from seeing so many foul and shabby buildings, especially in the heavily ethnic and horribly impoverished lower

Opposition to the el was well organized. This print from a pamphlet, widely circulated in the city to advertise the horrors of the elevated railway, shows stampeding horses, trampled children, collision, billowing black smoke, and searing steam. COURTESY THE NEW YORK HISTORICAL SOCIETY.

East Side. The Elevated Snake, as it was sometimes called, intruded into some of the loveliest parks and squares of the city. It slithered arrogantly through the middle of Battery Park en route from South Ferry—sanctuary in summer for hundreds of wage earners fleeing their hot tenements for a breath of cool air. For years there was agitation to uproot the interloper from this choicest of green parks, but the tracks stayed until the very end. Its invasion of the quiet privacy of Bryant Park, City Hall Green, tiny Jeanette Park, and lower Morningside Park were scarcely more merciful. And Hanover, Chatham, and Herald Squares were never the same after 1878.

The personal irritations brought to residents along the route of the els were manifold. *Scientific American* complained that the locomotives burned a poor grade of coal—one rich in sulphur—with the result that dwellers along the line as well as pedestrians were often nauseated by the stench. Said one editor, "it is like putting a foul chimney in front of everyone's bedroom window." For a time there were a good many stories in the newspapers about store awnings set ablaze from sparks from passing locomotives, but these minor conflagrations disappeared rapidly as the companies improved the spark arresters on their stacks. Some reporters wrote that ashes, hot water, and oil dripped into the streets and sidewalks, but these accounts were probably exaggerated, because the companies seemed extraordinarily careful about the quality of their anthracite and the reliability of the drip pans under their engines. One journal despaired about the peculiar rash of optical ailments that plagued the city in the 1880s blaming the elevated lines for throwing iron dust into the eyes of pedestrians traveling under the structure. Since the trains ran at high speeds and required strong stopping power, the brake shoes ground off fine showers of jagged, iron splinters that flew in all directions into the streets and sidewalks. Allegedly the awnings of shops along the route were positively rust-red with the stain of this metallic shower. *Scientific American* called for the invention (such as perhaps a huge magnet) that would prevent this "evil" that had sent, they said, hundreds of people to hospitals.

Of all the uproar caused by the elevated roads the worst seems to have been the noise—the deafening, bewildering racket of countless trains

The gaunt trestlework brought twilight to miles of street. It darkened the lower stories of many homes and stores, and the shade kept pavement and sidewalk damp and wet long after others were dry. COURTESY THE SMITHSONIAN INSTITUTION.

thundering along the overhead iron bridge, every sound and vibration intensified and reflected downward by the huge sounding board of the structure and the car bottoms. Along the lines the furious din caused by the pounding of the car wheels on the track was almost continuous. The roar of metal striking against metal was like a hammer on an anvil, the crash echoing and re-echoing through the hundreds of iron girders and braces and sweeping in undulating waves through the streets, ricocheting and reverberating through the brick and stone canyons of the city. The furious shriek and hiss caused by the trains must have strained the nerves of shopkeepers and families who desired to transact business, sleep, and hear themselves converse. When the Third Avenue line was opened past Cooper Union, the

noise was so appalling that a dozen classrooms had to be transferred to the Fifth Avenue side of the building. Abram Hewitt, secretary of the union, sent his friend and Gramercy Park neighbor Cyrus Field a bill for $540 to cover the cost of the move, protesting that it was unfair for the company to impede the education of the poor boys and girls of the school. Field gladly paid the bill.

Recently Dr. Susan Pettiss recalled a similar experience with the Third Avenue el. A social worker for UNRRA, Dr. Pettiss had been one of the first to enter Nazi concentration camps after the war. She survived that experience, but in 1946, upon taking a small apartment in Murray Hill just off Third Avenue, she came to grips with the el. "You would absolutely not believe the noise of those trains," she said. "Every time one would pass, my whole apartment would tremble. I just couldn't bear it." She moved at the end of the week. Few residents, however, were as mobile as Dr. Pettiss, because low rents held many poor families in their el-side flats.

No discussion of the dreadful din of the elevated would be complete without reference to William Dean Howells's eloquent *Letters of a Traveler from Alturia*. A utopian work, the tone of moral outrage against the railroads is particularly vehement. Of those who dwelt along the route of the trains, Howells had this to say: "People are born and married, and live and die in the midst of an uproar so frantic that you would think they would go mad of it; and I believe that physicians really attribute something of the growing prevalence of neurotic disorders to the wear and tear of the nerves from the vivid rush of the trains passing almost momently, and the perpetual jarring of earth and air from their swift transit. I once spent an evening in one of these apartments, which a friend had taken for a few weeks last spring . . . and as the weather had begun to be warm, we had the windows open, and so we had the full effect of the railroad operated under them. My friend had become accustomed to it, but for me it was an affliction which I cannot give you any notion of. The trains seemed to be in the room with us, and I sat as if I had a locomotive in my lap. Their shrieks and groans burst every sentence I began. . . . I cannot tell you how this brutal clamor insulted me, and made the mere exchange of thought a part of the squalid struggle. . . . I came away after a few hours of it, bewildered and bruised, as if I had been beaten upon with hammers. Some of the apartments on the elevated lines are very good, as such things go . . . but most of them belong to people who must dwell in them summer and winter, for want of money . . . and who must suffer incessantly from the noise I could not bear for a few hours. In health it is bad enough, but in sickness it must be horrible beyond all parallel. Imagine a mother with a dying child in such a place; or a wife bending over the pillow of her husband to catch the last faint whisper of farewell, as a Harlem train of five or six cars goes roaring by the open window! What horror, what profanation!"

Of all the uproar caused by the elevated, the worst, undoubtedly, was the furious din it created. The deafening continuous racket of the trains strained the nerves of those shopkeepers and families who attempted to conduct business, sleep, or simply hear themselves converse. COURTESY PRINTS AND PHOTOS DIVISION, LIBRARY OF CONGRESS.

In spite of all the noise and disorder, agitation and lawsuit, the elevated roads prospered. People adjusted to them the way a man adjusts to a chronic disease. The noise gradually blended into the general din of the city. No doubt the el was very noisy, but so was the city, the streets always throbbing with the clatter of drays and delivery carts, horsecars, coaches, and carriages of all sorts, and the sharp clip-clop of iron-shod hooves on the rough Belgian block paving. Later the endless parade of autos, trucks, and buses filling the avenues and the shrill jet whine of airplanes high overhead is no less loud. In the old days the dirt and noise of the trains were not so much noticed as they would be today, so conscious are we of environmental pollution. New York was,

and still is for that matter, a dirty city, especially during nineteenth-century winters, what with the tens of thousands of coal and wood fires pouring carbon into the atmosphere. People then accepted the tangential ills of progress—in rapid transit as well as in other areas of life. The horrible forest of tall poles that infested New York streets, carrying electric and telephone wires to their homes, was far from aesthetic. Also, the elevated did not pass through the more fashionable districts, such as Fifth Avenue, where influential residents lived in brownstone dignity.

Although almost everybody disliked the overhead railways at first, the disagreeable impression they created lessened with familiarity, and they came to be regarded as a necessary evil. The key to their eventual popularity was struck in the following playful words of a writer for the *American Architect* of 1883. After exhausting the vocabulary of his scorn and dislike upon the elevated roads for

The encroachment of the el was severe—almost total—in some streets. A Second Avenue train on Division Street coming into Chatham Square. C. 1890.
COURTESY THE COLLECTION OF E. ALFRED SEIBEL.

The great elevated snake, as it was sometimes called, intruded into some of the loveliest parks and squares in the city. Here it slithers through tiny Jeanette Park on the lower East Side, nearly grazing the windows of crowded tenements.

became the scene of a building boom, attracting thousands of new residents. Until the building of the Sixth Avenue elevated there was practically no way for a man of moderate means to reach his home in that section of the city, since surface lines were too slow. The el was the first agency to open up the area to population and gave it the first impetus to building and development. Armies of carpenters and brickmasons began rearing blocks of dwellings and apartments where just a few years before was nothing but barren rocky fields and rickety shanties. This West Side district had many natural advantages but was left to an unproductive and unprofitable existence until the railroads tapped its resources for population. Families who had been forced to live in Brooklyn or the New Jersey suburbs began returning to the city, and in a few years the fine old country seats on the West Side as far up as Washington Heights gave way to solid blocks of brick and brownstone, since with efficient transportation the land became too valuable to be used for lawns and gardens. By the turn of the century the whole upper part of the island was as closely developed as the districts below the park. It was hoped that the elevated railroads would provide some sort of relief for the tenement dwellers in the vast rookeries of the lower East Side, allowing them to move out of the congested slums into the more salubrious air of Harlem, where the rents were cheap. Unfortunately, those desperate people seemed to be so locked into their squalid ghettos that it took more than rapid transit to pry them out. And the places left by those fortunate enough to escape were quickly filled by the cargo of the next immigrant ship.

their damaging effect on certain examples of noble architecture he concluded by saying: "But let us shake off the general dustiness we have gathered by our walk along the substructure of the road, forget the holes burned in our coats by hot cinders, overlook the few grease-splashes upon our summer hat, and forgive the brakeman who found amusement in squirting tobacco juice down upon us, and let us go up and follow the unthinking populace in encouraging the monopoly. Oh, how delightful! Bless me, here we are at old Trinity again! I take it all back; let architecture and property rights and personal privilege and past associations perish, so long as we can fly through the air without following suit."

Clearly, there were obvious drawbacks to the elevateds, but on the whole they were a great boon for the city—especially to the real-estate speculators who made a killing from selling land in the areas the els opened up for development when the upper reaches of Manhattan Island were brought within rapid and easy reach of the business quarters downtown. Because of the el the upper West Side of the city along Central Park and above

For the working classes of the city the elevateds were an immense improvement in comfort and speed over the crowded, miserable horsecars, cutting their daily ride to and from their jobs by half. In good weather and with a clear track the horsecars took from three quarters of an hour to fifty minutes from Fifty-ninth Street to City Hall. The elevated trains made the same distance in twenty-eight minutes. As a result the trains became extraordinarily popular, and there was a sharp change in the general attitude toward the el, away from the bitter denunciation of just a few years earlier into general praise for its efficiency and public service. *Scientific American* did an about face from their stance as the el's severest critic when in 1883 they admitted that no other city

The tracks of the elevated spoiled the beauty of many fine buildings, insolently drawing its thick black line across their facades. This mansarded and towered edifice on Third Avenue at Sixty-fifth Street was designed by Henry Hardenbergh, architect of the old Waldorf-Astoria Hotel. COURTESY THE SMITHSONIAN INSTITUTION.

in the world had such an admirable transit system as New York's four lines of elevated railroad. The *National Car Builder* that same year spoke rapturously about the service, "the speed of the trains, the capacious, easy riding cars, well warmed and lighted, freed from obstructions, comfortable stations, and waiting rooms with gate and platform men charged with duties conducive to the safety and convenience of passengers." *King's Handbook of New York City* (1893), a sort of Chamber of Commerce auxiliary, was understandably quite enthusiastic in lauding the elevated, calling the system a "crowning achievement" in solving the

problems of rapid transit," for New Yorkers could fly through the air from end to end of their teeming island at railway speed in gorgeous Pullman cars. Instead of being dragged through the darkness and poisoned air of subways, passengers were borne swiftly high up above the crowded streets in the fresh, open air while watching the wonderful changing panorama of the Empire City through their windows. Even the stations were in midair, King pointed out, so that the ride from stop to stop was like the airy swiftness of the swallow's flight.

The elevated was far more than efficient transporation for most people who rode it. It was adventure. It was romance. It was unfailing entertainment from a grandstand seat. *The Hazard of New Fortunes* (1890), the most immediately successful of William Dean Howells's novels, was written not long after he moved from Boston to New York to join the editorial staff of *Harpers* magazine. The book reflects the impact of that change, because the main character, Basil March, has also made a similar move to edit a magazine. In the course of the book are several accounts of travel on the elevated; these passages are an interesting mixture of the best and the worst about the roads—the horror of living along the route and the fascination and pure joy of riding on the trains: "perfectly atrocious, of course," he says, "but incomparably picturesque!" He seems to be saying that the elevated was like the city itself, a monstrous invasion to privacy but wonderfully alive and gay. One passage in the novel is unsurpassed in capturing the spirit of romance and adventure that made riding the trains so extraordinary. "At Third Avenue they took the Elevated, for which she confessed an infatuation. She declared it the most ideal way of getting about in the world, and was not ashamed when he reminded

her of how she used to say that nothing under the sun could induce her to travel on it. She now said that the night transit was even more interesting than the day, and that the fleeting intimacy you formed with people in second and third floor interiors, while all the usual street life went on underneath, had a domestic intensity mixed with a perfect repose that was the last effect of a good society with all its security and exclusiveness. He said it was better than the theatre, of which it reminded him, to see those people through their windows: a family party of work-folk at a late tea, some of the men in their shirt-sleeves; a woman sewing by a lamp; a mother laying her child in its cradle; a man with his head fallen on his hands upon a table; a girl and her lover leaning over the window-sill together. What suggestion! What drama! What infinite interest! At the Forty-second Street station they stopped a minute on to the

Sprawling hugely over Forty-second Street, the spur to Grand Central Station.

bridge that crosses the track to the branch road for the Central Depot, and looked up and down the long stretch of the Elevated to north and south. The track that found and lost itself a thousand times in the flare and tremor of the innumerable lights; the moony sheen of the electrics mixing with the reddish points and blots of gas far and near; the architectural shapes of houses and churches and towers rescued by the obscurity from all that was ignoble in them, and the coming and going of the trains marking the stations with vivider or fainter plumes of flame-shot steam— formed an incomparable perspective. They often talked afterward of the superb spectacle."

13 Operations, Traffic, and Safety

The practical operation of an elevated railway was unique—far different from the conditions a train crew would encounter on a conventional surface railroad. For one reason the route was over a continuous bridge with the height from the rails to the ground varying from twelve to fifty-seven feet. Along a good portion of the road the engineer or motorman had to stop every forty seconds to drop off and take on passengers; it meant starting up again and again, accelerating, shutting off, and stopping at a chalk mark within the interval prescribed by the timetable. Stops varied from three seconds upward, with an average of ten seconds, depending on the hour of day. During rush hours the locomotives or motor cars pulled five or more coaches filled to the doors with people, and the headway between trains was counted in seconds, not minutes. Regular running time allowed less than a minute between trains; from midnight to sunrise intervals were from seven to fifteen minutes. The trip from the upper and lower terminus, about nine miles, was run in about thirty-five minutes, including twenty-five stops. Express trains were run on some lines by using a third track in the center, and since most traffic was downtown in the morning and uptown at night, the space could be used to store extra cars in between.

The gateman usually called out the destination of each train as it approached the station during steam years, though most people could tell the routing from the signals on the locomotive as it approached. Since elevated trains had different destinations even on the same line, marker light signals indicated the route. Before electrification these lights were mounted on the locomotive car roofs and illuminated at night by kerosene lamps. Afterward all motor cars were so equipped at each

The first serious accident involving an elevated railroad train occurred in 1879, just when the citywide system was getting started. As a result, the accident was damaging to the publicity-conscious company, as a shock wave ran through the city about the careless operations on the el. On March 25, two trains on the Third Avenue line crashed head on at a turnout near the Forty-second Street station. They collided with such momentum that the locomotives were lifted into the air. Several passengers were injured, two rather seriously, though none died in the mishap. Cause of the wreck was a switch left open by a careless switchman. COURTESY LIBRARY OF CONGRESS.

end beside the headlight. The four sides of the marker were white, red, green, and yellow, and the lamp could be turned by a spindle extending through the car roof to indicate the type of service, be it local, express, local express, through express, shuttle, or rush-hour extra. Local expresses operated express south of 149th Street, while through expresses operated south of 177th Street; north of those points they operated as locals, making all stops. During the 1880s and early 1890s the Second and Ninth Avenue elevated lines (the least busy of the els) were closed at night from 8 P.M. to 5:30 A.M. and all day on Sunday. According to the *Times*, local residents were heartily thankful for the rest on Sunday: "They attend church less than they have done, staying home to enjoy the comparative quiet and to realize wholly their deliverance from the infernal trains." Many

el-side dwellers used the day to sleep, because they could not during the week.

A ride on the el at first cost ten cents, except during morning and evening rush hours, when the fare was dropped to five cents for the benefit of the workers of the city. Then in 1883 Sunday fares were reduced to five cents. In 1886, to encourage patronage of the least popular routes, the company cut the fare all day on the Second Avenue line and the Ninth Avenue line below Fifty-third Street; later that same year fares were cut across the board, and a nickel bought a ride on all Manhattan elevated trains all day. This five-cent fare remained in effect for forty years until 1946, when postwar inflation pushed it up to ten again. At first tickets were sold at the stations and collected on the trains by conductors, but as traffic increased, this system caused considerable confusion, because some passengers managed to keep their tickets on short rides by moving as far as they could from the collector. Consequently, in 1880 the company installed cancelling boxes into which passengers would drop their tickets before entering the train. The coin-operated turnstiles used in the last years of the elevated came into use in 1923.

Rather little in the way of special service appeared on the elevated after the first flirtation with first-class saloon cars about 1878, the short-lived bicycle cars in the mid-nineties, and a few open summer cars before World War I. There is a story that George Pullman, whose company had built many cars for the elevated, chartered a special train one June day to take him and his entourage downtown to board an ocean liner bound for Europe, but the story is not substantiated and is likely to be apocryphal. What few people realize, however, is that beginning about 1907 the els operated a freight and express service under contract with Wells Fargo. Five trains a day loaded with merchandise and livestock from Erie's freight piers on the Hudson River rumbled over the high rails between special depots at 39 Greenwich Street (later the site of the Cunard Building), 129th Street between Second and Third avenues, and at 53rd Street at Eighth Avenue. Hefty hydraulic elevators boosted pianos, bananas, chickens, and cattle up to the elevated trains at one end of the route and lowered them at the other end, where waiting express wagons and trucks picked up the freight for delivery in Harlem and the Bronx. Most of the business was con-

ducted late at night when the system was at its quietest. For many years the elevated also carried mailbags on the front platform of many regular trains. The company apparently carried mail as a public service; since they earned only $5,000 in most years, it could hardly be called a lucrative operation. Both the mail and freight service were discontinued shortly before World War I.

The successful operation of the elevated was largely dependent on its army of employees. And it was in fact very much like an army: with a heirarchy of managers called officers and rank upon rank of soldiers who were conductors, ticket-takers, firemen, and repairmen. The number and frequency of trains and the enormous number of passengers (500,000 daily in 1893) required nothing but the highest grade of service. To accomplish this job, Manhattan Elevated and its successor, the IRT Elevated Division, needed more employees than the entire New Haven Railroad and nearly ten times as many men per mile than the New York Central. In some of its departments the elevated had a list of spare or supernumerary men larger than the regular force

Three people were killed in this wreck when a Bronx Park local telescoped an empty train at the 175th Street station on the Third Avenue el in 1919.

Disaster struck the elevated in 1905 when a speeding rush-hour train left the track at the Fifty-third Street curve. The momentum was so great that one car turned a complete loop in the air before crashing into the street. Another car ran over the edge of the structure and hung suspended halfway out in space. Thirteen passengers were killed and forty-eight seriously injured. COURTESY THE COLLECTION OF E. ALFRED SEIBEL.

on many roads of considerable importance. In 1881 they employed 3,274 men; in 1887, 4,586; in 1893, 5,000; and in 1931, nearly 6,000.

The company's employees were divided into seven departments: The Engineer's Office, Department of Building Repairs, Roadmaster's Department, Department of Street Repairs, General Ticket Department, Transportation Department, and the Mechanical Department. The executive management of the elevated after the Manhattan Railway takeover in 1880 was largely in the hands of Gould's chief aide, Col. Frank Hain, general

manager and later second vice-president of the company. President of the road until his death in 1892, though not intimately involved in the everyday operation of the company, was Jay Gould, after which his eldest son, George J. Gould, became head. Other Gould sons, Edwin and Howard, served on the board of directors of the largely family-controlled corporation. Cyrus Field was an active junior partner for many years as was Russell Sage, who was prominent in the executive management of the road until the lines were leased to the IRT in 1903.

Largest in terms of employees was the Mechanical Department (1,814 men in 1887), which was responsible for design, acquisition, and repair of all rolling stock. The body was composed of the master mechanic, blacksmiths, boilermakers, carpenters, car cleaners and car inspectors, coal tenders, engine dispatchers, engine inspectors, stationary and locomotive engineers, firemen, engine wipers, lamp trimmers, machinists, painters, pipe felters, and water tenders.

The Transportation Department (employing 1,322 men in 1887) was under the immediate charge of the superintendent, who with his lieutenants, the train masters, literally held the threads of the thousands of details of running the trains. They arranged the timetables and coordinated the operation of 2,500 daily elevated trains on Manhattan Island alone. (The total number of trains for the month of May 1887 was 76,804 on the city's main lines.) On Saturday and Sundays an entirely new timetable had to be worked out, and the same was true of holidays, when additional trains had to be added to the whole system. The intricacies and difficulties of an ordinary elevated schedule resulted from the need for starting trains from the termini so that they fitted in properly with those trains already started from intermediate stations, the constant aim being to have the intervals between trains on any given portion of the route always as nearly uniform as possible. Trains also had to be scheduled so that the crew would work no more than the legally prescribed ten hours a day and still have an hour to eat their lunch or dinner, at the end of a trip preferably and at a proper time of day. And even the most intricate schedule could be wrecked by a breakdown of one train on the line. The Transportation Department was headed by the superintendent and trainmasters, who supervised the work of the conductors and guards who rode toward the trains. Head of the train crew was the conductor, who rode in the head car and supervised the guards in other cars of the train. Other key men of the department were dispatchers, car couplers, flagmen, switchmen (hand and tower), and the telegraph operators.

All the station employees were placed in the General Ticket Department, working under the general ticket agent. Because there were 174 separate elevated stations, the company needed a large crew of station agents (311). Since there were usually two separate stations at each stop (northbound and southbound), separate crews were required to man each facility, so that the station force seemed quite large. Other employees of the department were the gatemen, platform men, porters, station inspectors (four day and three night), and accountants and clerks. New men applying for station work were examined by the

company's surgeon and by the general manager's office for mental capacity. Application blanks for employment stated that a man had to be over twenty-one and under forty-five years old, at least five feet six inches in height, able to read and write English, and sane and sound of body. He also had to be familiar with the streets and streetcar lines of New York City, its points of interest, and the terminals and routes of surface railroads and ferries connecting with the elevated. All men accepted for work on the el started first as gatemen or car cleaners, from which they might be eligible for promotion to the ranks of platform men, station agents, guards, and firemen. The force of gatemen was the recruiting ground for station agents and trainmen, by gradually advancing and testing good men in a natural manner. No man was employed as a guard, for example, until he had worked for a season or two at a station where he was regularly watched and where he could, if he were ambitious, learn much of the trainman's work by observation. The elevated administered its discipline in much the same way as other railroads. Punishments were in the form of suspensions or discharge; admonition was embodied in circulars posted in the offices where men reported for duty. In some cases the men were sent letters by inspectors, who daily were about their work in the stations and on the line.

The Roadmaster's Department (thirty-six men) was responsible for the maintenance and repair of track and structure. During the busy day and night schedules and without halting the trains, gangs of track repairmen renewed rails or replaced switches, sometimes in as short as three-and-a-half minutes (which was about the maximum time between trains during the day). Of the eighteen-man gangs, eight men would lift the rails in and out; four would pull spikes; two adjust the joint fastenings; and four would respike. Their coordination had to be perfect to complete the work before the arrival of the next train just minutes off. Consequently, men who were too old to learn or too young to be cautious or whose health or habits were likely to cause trouble were rejected from the work. The Roadmaster's Office was open day and night; if an interlocking machine went out on any part of the system, the yardman had to be prepared to act without delay, and until the apparatus was repaired a trackman had to act as a flag-signaler for the towerman.

The department had fourteen wrecking outfits

A coaling station in lower Greenwich Street serviced both Ninth and Sixth Avenue el trains, whose tracks converged here and ran on joint structures through Battery Park and South Ferry. COURTESY THE NEW YORK HISTORICAL SOCIETY.

Manhattan Elevated engine no. 60 at the Third Avenue line's Eighty-ninth Street station, March 11, 1902. This was one of the last runs of a steam locomotive on the system. Electrification was but a few months off. COURTESY THE COLLECTION OF E. ALFRED SEIBEL.

besides a large amount of materials and tools for use on the street level in changing and repairing foundations. The structure division was an important unit of the Roadmaster's Department because building and excavating that might endanger the foundations of the columns were going on somewhere along the elevated route at all times. Six inspectors constantly patrolled the road, keeping watch over any adjacent building operations as well as attending to their regular duty of scrutinizing the ironwork for weakening or disrepair caused by the immense traffic, whose jarring and pounding was enough to shake rivets loose. (On the Third Avenue line alone there were two and a half million rivets to watch.) The thirteen pipe fitters in the department were engaged in keeping the five hundred drip pans on the structure in good order. These eight-by-forty-foot pans prevented the dripping of water and oil into the street from passing engines and cars.

The twenty-five men in the Department of Street Repairs worked at putting up and taking down temporary trestles and resetting foundations removed or changed because of construction work in the streets along the route. The men in the department worked as pavers, masons, laborers, under the foreman of street repairs. A separate Department of Building Repairs employing 74 men saw to the care of all elevated buildings and stations. The Engineer's Office employed a chief engineer, two assistant engineers, and five aides.

The men who worked for the elevated were paid moderately well for the times, and the jobs were secure and much sought after. The great number of employees gave rise to a vast patronage system as company officials (Manhattan as well as IRT) gave jobs to party loyalists in return from favors from City Hall. The el had many jobs to give out, particularly at the unskilled level, and so it seemed only natural to use employment as a vehicle for political emolument. Highest paid of the rank-and-file workers were the locomotive engineers and later the motormen. In the 1880s they received from $3.00 to $3.50 a day, depending on years of service. By 1894 they received $100 a month. Ticket agents received from $1.75 to $2.25 a day; conductors $1.90 to $2.50; guard and trackmen $1.50 to $1.65; and gatemen from $1.25 to $1.50 per day. By the 1930s most wages on the elevated had nearly doubled, but then so had costs and prices. Many of the more skilled men, especially the locomotive engineers and firemen, were unionized in the late 1880s, though Jay Gould had no use at all for the union movement or their leaders and was often very difficult when they approached him for negotiations. When they complained of having to work such long hours for a daily wage of $1.75 to $2.00, Gould listened courteously to their grievances and agreed to the principle of a ten-hour day. Their gratitude for cutting the workday was, however, short lived, for he had simultaneously ordered that their wages be cut accordingly.

Gould was not really as antisocial as he was often accused of being; he was merely antiunion. Under his aegis the company published a magazine called the *Elevated Railroad Journal,* especially for his workers. Actually the *Journal* was a quasiofficial organ of the company, since the words "Manhattan Railway, Jay Gould President" appeared as a sort of submasthead at the top of the cover. Despite its official-sounding name the publication was not a professional journal full of significant, inside details on the operations of the

elevated railroads. It was, rather, a very low-key magazine, popular and sentimental in tone and chatty and highly informal in style with typically Victorian poems, letters, and articles titled, for example, "Too Late" or "Wrestling Camels." The pages carried general advertisements for ladies' clothing, fire insurance, and some bits and pieces of railroad gear, but there was very little elevated railroad information at all. The *Elevated Railroad Journal* is now an obscure and forgotten publication, appearing neither in the Library of Congress serials catalog nor the Union List. Fortunately, one brittle, old copy is preserved in that golden hoard of the New York Public Library.

Employees of Manhattan Elevated published on their own a very handy guidebook to help residents and tourists get around the city on their trains. Called the *Train Men's New York City Guide Via Elevated Railroads,* the little booklet went through nine editions by 1889. Two editions were published yearly, spring and fall, by editor F. B. Knowlton, who worked at the el station at 129th Street and Third Avenue. The guide contained valuable information for patrons, enabling them to find the proper line and stop for any street address in the city. It also gave the nearest el station for almost all points of interest in New York —amusements, banks, churches, hospitals, hotels, libraries, depots, etc. To find the Home for Destitute Girls (200 West 14th Street), for example, one went to the 14th Street station on the Sixth Avenue line. For the British Consulate (22 State Street) one went to South Ferry station.

Patronage on the elevated railways rose yearly despite depression, war, and competition from new types of transport, such as the cable cars that had a spurt of fashion in the late 1880s. There was a sharp rise in the tide of elevated travel in 1879 and 1880 because of the completion of the Sixth, Second, and Third Avenue lines. This was followed by a general though more gradual increase through the 1880s and early 1890s. Some erosion occured in the mid 1890s, no doubt resulting from the panic of 1893, which ushered in the worst depression the nation had experienced up to that time. By mid 1894 there were 192 railroads in receivership. In the first years of the new century the elevated was once again gaining steadily in patronage. Even after the subways got going in 1904 the number of people who rode the elevated trains continued to grow hugely. By the mid 1930s, what with the national depression and the abandonment of some el service, patronage fell off sharply, and the subways were hauling six times as many passengers. The following table illustrates some of the basic figures in the number of passengers carried on the elevated. For additional statistical information the reader should consult *Poors Manual* (up to 1912), the New York State Railway Commission Reports, the New York State Transit Commission Reports, and the New York State Public Service Commission Reports.

Number of Passengers Carried on Elevated Trains in New York City

Year	Passengers	Year	Passengers
1872	242,000	1897	183,000,000
1873	640,000	1899	174,000,000
1875	910,000	1902	215,000,000
1876	2,000,000	1903	246,000,000
1878	4,000,000	1905	266,000,000
1879	14,000,000	1907	283,000,000
1880	60,000,000	1910	294,000,000
1883	93,000,000		
1885	103,000,000	1921	374,000,000
1886	115,000,000		
1837	159,000,000	1931	327,000,000
1888	172,000,000		
1890	186,000,000	1938	200,000,000
1893	221,000,000	1939	169,414,000
1895	188,000,000		

Heaviest traffic was on the Third Avenue and Sixth Avenue lines, which ran through the commercial and resident heartland of the city; each carried about a thousand trains a day in 1892. Least crowded of the el was the one on Ninth Avenue, which ran only about 400 trains daily. Typical of other city transit systems, the elevated's biggest passenger month was December during the Christmas rush; slowest were July and August, traditional vacation time for New Yorkers escaping the searing city heat.

The operations of the Manhattan Railway were immensely profitable for the owners, since for years the elevated held a legal monopoly on rapid transit in the superpopulous Empire City. Until the building of the first subway and public takeover of the roads by IRT in 1904, the yearly revenue collected by Jay Gould and his heirs (even with the low five-cent fare) amounted to a positive mountain of nickels. Net earnings for the company for 1880 was just short of $2,000,000; more than $3,000,000 by 1886; and $4,000,000 by 1890. Compared with other passenger railroads in the United States, Manhattan was far and away the leader in net earnings per mile of road. The following table illustrates the disparity between elevated earnings and those of other railroads. Note that the figures have been chosen from the more affluent roads in the country, not from small, branch lines. Actually the average net earnings per mile in the country during the decade of the 1880s were only $2,560.

Net Earnings Per Mile of Railroad

Railroad	1881	1885	1887
Pennsylvania Railroad	10,386.00	7,641.00	8,152.00
New York Central	10,635.00	6,398.00	7,457.00
Union Pacific	4,201.00	2,165.00	2,356.00
Baltimore & Ohio	8,151.00	6,616.00	7,239.00
Manhattan Elevated	58,902.00	87,342.00	111,240.00

The West Side elevated lines on Manhattan Island ended at the Harlem River. Here the company built their 159th Street yards for storing locomotives and cars (mid left center). The Polo Grounds, early home of the New York Giants, was just next door. In the background is High Bridge, an aqueduct built in 1848 to bring the city Croton Reservoir water from the mainland. COURTESY THE COLLECTION OF E. ALFRED SEIBEL.

Railroading has been a hazardous business for passengers as well as for employees since the first primitive trains began running in the 1830s. Locomotive boilers burst by the score over the years, dismembering engineers and firemen; coaches derailed, mangling passengers in the shattered debris; rear-end collisions caused flimsy, wooden cars to telescope, impaling women and children in the splintered wreckage; and bridges and viaducts collapsed, tumbling heavy trains to watery death. The columns of the popular weekly journals such as *Harpers* and *Leslies* flowed with the bloody dangers of rail travel during the nineteenth century, and the grim statistics of

mortality continued far into the new century. When the pioneer Ninth Avenue elevated railway was opened in 1870, pessimists quite naturally predicted that the cars would constantly be falling off the track into the streets. Journalists and engineers were also quick to point out the folly of such transit, saying that the fundamental risk of riding on tracks so high in the air was tantamount to disaster. The continuous light trestle structure of the elevated, supported by slender iron columns from forty to sixty feet apart was, they insisted, at any time liable to collapse from the weight of the heavily loaded trains. The fiercely congested traffic and short headway between trains, especially during the rush hours, caused widespread fear of rear end collision. Just the slightest mishap from a broken rail, wheel, or axle would throw a train of cars down into the street with almost certain mass death. And the overtaxed structure itself was constantly exposed to vandalism, accident, or demolition. Even a fire in an adjacent building could tumble down high masonry walls

MANHATTAN RAILWAY.

OPERATING ALL ELEVATED LINES IN NEW YORK CITY.

GEO. J. GOULD, President.   JOHN WATERHOUSE, Chief Engineer.   CHAS. P. McFADDIN, Gen. Ticket Agent.
——, 1st Vice-President.   D. W. McWILLIAMS, Sec'y and Treasurer.   General Offices—No. 71 Broadway, N. Y.
F. K. HAIN, 2d Vice-Prest. & Gen. Man'r.   E. F. J. GAYNOR, Auditor.

EASTERN DIVISION.

SECOND AVENUE LINE.

Mls.	Minutes.	STATIONS.	LOCATION.
.0	0	South Ferry [1]	Foot Whitehall Street.
.39	2	Hanover Square [15]	Pearl and So. William Sts.
.74	4	Fulton Street [14]	Pearl and Fulton Streets.
.93	5	Franklin Square	Pearl and Cherry Streets.
1.32	7	Chatham Junction	Pearl and Bowery.
1.63	9	Canal Street *	Allen and Canal Streets.
1.80	10	Grand Street *	Allen and Grand Streets.
2.03	11	Rivington Street	Allen and Rivington Sts.
2.23	12	First Street	First Street and 1st Ave.
2.57	13½	Eighth Street *	Eighth Street and 1st Ave.
2.87	15	Fourteenth Street *	14th Street and 1st Ave.
3.11	16	Nineteenth Street	19th Street and 1st Ave.
3.39	17½	Twenty-third Street *	23d St., bet. 1st and 2d Av.
3.99	20	Thirty-fourth Street [2]	34th Street and 2d Ave.
4.38	21½	Forty-second Street	42d Street and 2d Ave.
4.78	23	Fiftieth Street	50th Street and 2d Ave.
5.13	24½	Fifty-seventh Street	57th Street and 2d Ave.
5.53	26½	Sixty-fifth Street	65th Street and 2d Ave.
6.30	29½	Eightieth Street	80th Street and 2d Ave.
6.60	31	Eighty-sixth Street	86th Street and 2d Ave.
6.90	32½	Ninety-second Street	92d Street and 2d Ave.
7.18	34½	Ninety-ninth Street	99th Street and 2d Ave.
7.85	38	One Hundred and Eleventh Street	111th Street and 2d Ave.
8.10	39½	One Hundred and Seventeenth St.	117th Street and 2d Ave.
8.30	40½	One Hundred and Twenty-first St	121st Street and 2d Ave.
8.66	43	One Hundred and Twenty-seventh Street	127th Street and 2d Ave.
8.76	43½	One Hundred and Twenty-ninth Street [11]	129th Street and 2d Ave.

Daily trains at intervals of 2 to 6 minutes between 4 43 a.m. and midnight. Line closed between midnight and 4 43 a.m.

THIRD AVENUE LINE.

Mls.	Minutes.	STATIONS.	LOCATION.
.0	0	South Ferry [1]	Foot Whitehall Street.
.39	2	Hanover Square [15]	Pearl and So. William Sts.
.74	4	Fulton Street [14]	Pearl and Fulton Streets.
.93	5	Franklin Square	Pearl and Cherry Streets.
*		City Hall [3]	Chatham and Centre Sts.
1.32	7	Chatham Square	Chatham St. and Bowery.
1.46	7½	Canal Street *	Bowery and Canal Street.
1.64	8½	Grand Street *	Bowery and Grand St.
2.04	11	Houston Street *	Bowery and Houston St.
2.49	13½	Ninth Street *	3d Ave. and Ninth Street.
2.74	15	Fourteenth Street *	3d Ave. and 14th Street.
2.94	16	Eighteenth Street *	3d Ave. and 18th Street.
3.18	17½	Twenty-third Street *	3d Ave. and 23d Street.
3.42	19	Twenty-eighth Street	3d Ave. and 28th Street.
3.72	20½	Thirty-fourth Street [2]	3d Ave. and 34th Street.
		Thirty-fourth Street Ferry [2]	Foot 34th Street, E.R. [2]
4.11	23	Forty-second Street [4]	3d Ave. and 42d Street.
		Grand Central [4]	42d Street and 4th Ave.
4.30	24	Forty-seventh Street	3d Ave. and 47th Street.
4.66	25½	Fifty-third Street	3d Ave. and 53d Street.
4.96	27	Fifty-ninth Street [16] *	3d Ave. and 59th Street.
5.35	29	Sixty-seventh Street	3d Ave. and 67th Street.
5.81	31	Seventy-sixth Street	3d Ave. and 76th Street.
6.22	33	Eighty-fourth Street	3d Ave. and 84th Street.
6.47	34	Eighty-ninth Street	3d Ave. and 89th Street.
6.92	36½	Ninety-eighth Street	3d Ave. and 98th Street.
7.33	38	One Hundred and Sixth Street	3d Ave. and 106th Street.
7.83	40	One Hundred and Sixteenth Street	3d Ave. and 116th Street.
8.28	42	One Hundred and Twenty-fifth St.	3d Ave. and 125th Street.
8.48	43	One Hundred and Twenty-ninth St.	3d Ave. and 129th Street

Daily trains at intervals of 1 to 5 minutes between 4 45 a.m. and midnight, and at intervals of 15 minutes between midnight and 4 45 a.m.

CITY HALL BRANCH.—(City Hall to Chatham Square, 6.36 m.)—Daily trains every 15 minutes from midnight to 5 30 a.m., after which main line trains run to City Hall at intervals of 3 to 6 minutes.

34th STREET BRANCH.—(3d Ave. to East River, 0.39 m.)—Daily trains between 5 30 a.m. and midnight at intervals of 3 to 6 min.

42d STREET BR.—(Grand Central Station to 3d Ave., 0.20 m.)—Daily trains between 6 00 a.m. and midnight at intervals of 3 to 5 min.

WESTERN DIVISION.

SIXTH AVENUE LINE.

Mls.	Minutes.		LOCATION.
.0	0	South Ferry [1]	Foot Whitehall Street.
.30	1½	Battery Place [12]	Battery Pl. & Greenwich St.
.51	4	Rector Street [5, 10]	N. Church and Rector Sts.
.72	5	Cortlandt Street [6]	N. Church & Cortlandt Sts.
.95	6	Park Place [9]	Church St. and Park Place.
1.17	7½	Chambers Street [8]	Hudson & Chambers Sts.
1.44	9	Franklin Street [7]	W. B'way and Frank'n Sts.
1.70	10½	Grand Street *	S. 5th Ave. and Grand St.
2.19	13	Bleecker Street *	S. 5th Ave. and Bleecker St.
2.70	16	Eighth Street *	6th Ave. and 8th Street.
3.00	17½	Fourteenth Street *	6th Ave. and 14th Street.
3.19	18½	Eighteenth Street	6th Ave. and 18th Street.
3.44	20	Twenty-third Street *	6th Ave. and 23d Street.
3.67	21	Twenty-eighth Street	6th Ave. and 28th Street.
3.93	22	Thirty-third Street *	6th Ave. and 33d Street.
4.38	24	Forty-second Street [4, 13]	6th Ave. and 42d Street.
4.78	26	Fiftieth Street	6th Ave. and 50th Street.
5.18	20	Fifty-eighth Street [16]	6th Ave. and 58th Street.
5.29	29	Fifty-third Street	8th Ave. and 53d Street.
5.77	33	Fifty-ninth Street *	9th Ave. and 59th Street.
6.11	34½	Sixty-sixth Street	9th Ave. and 66th Street.
6.41	35	Seventy-second Street	9th Ave. and 72d Street.
6.87	36½	Eighty-first Street.	9th Ave. and 81st Street.
7.47	38½	Ninety-third Street.	9th Ave. and 93d Street.
8.02	40	One Hundred and Fourth Street.	9th Ave. and 104th Street.
8.81	43½	One Hundred and Sixteenth St.	8th Ave. and 116th Street.
9.26	45½	One Hundred and Twenty-fifth St [4]	8th Ave. and 125th Street.
9.76	48	One Hundred and Thirty-fifth St.	8th Ave. and 135th Street.
10.26	50	One Hundred and Forty fifth St.	8th Ave. and 145th Street.
10.76	52	One Hundred and Fifty-fifth St. [10]	8th Ave. and 155th Street.

Daily trains at intervals of 2 to 4 minutes between 5 30 a.m. and midnight, and at intervals of 20 minutes between midnight and 5 30 a.m.

NINTH AVENUE LINE.

Mls.	Minutes.		LOCATION.
.0	0	South Ferry [1]	Foot Whitehall Street.
.30	1½	Battery Place [12]	Battery Pl. & Greenwich St.
.50	4	Rector Street [6]	Greenwich & Rector Sts.
.72	5	Cortlandt Street [6]	Greenwich & Cortl'dt Sts.
.92	6	Barclay Street [9]	Greenwich & Barclay Sts.
1.06	7	Warren Street [8]	Greenwich & Warren Sts.
1.32	9	Franklin Street [7]	Greenwich & Franklin Sts.
1.60	10	Desbrosses Street *	Greenwich & Desbro's Sts.
1.98	11½	Houston Street	Greenwich & Houst'n Sts.
2.26	12½	Christopher Street *	Greenwich & Christo'r Sts.
2.85	15	Fourteenth Street *	9th Ave. and 14th Street.
3.29	17	Twenty-third Street *	9th Ave. and 23d Street.
3.63	19	Thirtieth Street	9th Ave. and 30th Street.
3.83	20	Thirty-fourth Street	9th Ave. and 34th Street.
4.23	22	Forty-second Street [13]	9th Ave. and 42d Street.
4.63	24	Fiftieth Street	9th Ave. and 50th Street.
5.08	26	Fifty-ninth Street [16] *	9th Ave. and 59th Street.
5.42	27½	Sixty-sixth Street	9th Ave. and 66th Street.
5.72	28	Seventy-second Street	9th Ave. and 72d Street.
6.18	29½	Eighty first Street	9th Ave. and 81st Street.
6.78	31½	Ninety-third Street	9th Ave. and 93d Street.
7.33	33	One Hundred and Fourth Street	9th Ave. and 104th Street.
8.12	36½	One Hundred and Sixteenth St	8th Ave. and 116th Street
8.57	38½	One Hundred and Twenty-fifth St [4]	8th Ave. and 125th Street.
9.07	41	One Hundred and Thirty-fifth St.	8th Ave. and 135th Street.
9.57	43	One Hundred and Forty-fifth St.	8th Ave. and 145th Street.
10.07	45	One Hundred and Fifty-fifth St. [10]	8th Ave. and 155th Street.

Daily trains at intervals of 3 to 6 minutes between 5 33 a.m. and 7 57 p.m. Line closed between 8 00 p.m. and 5 30 a.m.

SUBURBAN BRANCH.

Mls.	Minutes.		LOCATION.
.0	0	129th St. and 3d Ave	3d Ave. and 129th Street.
.01	1	129th St. and 3d Ave	2d Ave. and 129th Street.
.46	3	One Hundred & Thirty-third St.	Between Alexander and Willis Avenues.
.67	4½	One Hundred & Thirty-eighth St	Between Alexander and Willis Avenues.
.92	6	One Hundred and Forty-third St.	
1.25	7½	One Hundred and Forty-ninth St.	3d Ave. and 149th Street.
1.66	9	One Hundred and Fifty-sixth St.	3d Ave. and 156th Street.
1.95	10	One Hundred and Sixty first St.	3d Ave. and 161st Street.
2.32	11½	One Hundred and Sixty-sixth St.	3d Ave. and 166th Street.
2.65	12½	One Hundred and Sixty-ninth St.	3d Ave. and 169th Street.
3.08	14	Wendover Avenue	3d Ave. & Wendover Ave.
3.39	15½	One Hundred and Seventy-fourth St.	3d Ave. and 174th Street.
3.71	17	One Hundred & Seventy-seventh St	3d Ave. and 177th Street.

Daily trains at intervals of 6 minutes between 5 08 a.m. and 11 15 p.m.; 10 to 15 minute intervals between 11 15 p.m. and 12 45 night. Line closed from 12 45 night to 5 08 a.m.

CONNECTIONS.—* Cross-town lines. [1] With Staten Island R.R., Manhattan Beach R.R., New York and Sea Beach R.R., and South and Hamilton Ferries for Brooklyn. [2] With 34th Street Ferry, and Long Island R.R. [3] With Brooklyn Bridge. [4] With New York Central & Hudson River; New York, New Haven & Hartford, and New York & Harlem R.Rs. [5] With New Jersey Southern R.R. [6] With Pennsylvania; Lehigh Valley; New York, Susquehanna & Western R.Rs.; Central R.R. of New Jersey, and New York & Long Branch R.R. [7] With West Shore and New York, Ontario & Western R.Rs. [8] With New York, Lake Erie & Western R.R. and Fall River Line. [9] With Delaware, Lackawanna & Western R.R. [10] With New York & Northern Ry., and trains for High Bridge. [11] With local trains New York, New Haven & Hartford R.R. for New Rochelle. [12] With Iron Steamboat Co. for Coney Island. [13] With West Shore and New York, Ontario & Western R.Rs. [14] With Fulton Ferry for Brooklyn. [15] With Wall Street Ferry for Brooklyn. [16] Central Park entrance.

onto the ironwork, which could but offer little resistance.

In the 1930s King Kong's rampage on the Third Avenue El made movie history. The giant ape ripped up the tracks just before the local rolled through. Then suddenly the motorman saw Kong and slammed on the brakes. As passengers were thrown together by the sudden halt, Kong's huge maddened eyes appear at the window of the El car. As women faint and strong men gasp, he picked up the car and threw it into the street. Fortunately, no such disaster ever happened to the El. In fact, despite the enormous potential for danger, the Elevated's safety record was nothing less than astonishing. The pessimists of 1870 were certainly poor prophets, for in the eighty years of their existence only one train ever fell into the street in Manhattan. (Two fell from the tracks on the neighboring Brooklyn el.) In the entire thirty-two years of steam operations locomotives once or twice plunged off the trestle, but no car ever followed. Even the old pioneer Greenwich Street el, for all its experiments, failed to spill any of its passengers into the sidewalk. Perhaps because of the inherent danger of elevated trains, the officers of the lines through the years were especially alert to the importance of safety, insisting on scrupulous inspection of rolling stock and constantly vigilant to the fragile nature of the structure, strained from the wear and tear of the constant and ever-increasing traffic.

Winter storms created difficult operational problems for the elevated management, since the open, exposed structure was particularly vulnerable to the season's cold, slashing wind. Snow was not usually much of a problem, because it was blown off or dropped between the ties, but ice occasionally formed in the trough between the track rail and the guard timber so that service had to be halted temporarily either until the ice melted or could be broken up by track gangs. Of all the winter storms to disrupt elevated service, worst by far was the famous Blizzard of 1888. Actually it was rather late for such a storm to strike the city—Sunday, March 11. And since most New Yorkers were preparing for spring (gardens were already greening and crocuses were blooming in Central Park), the idea of a blizzard seemed remote. Beginning first as heavy rain, the weather changed by teatime when a sudden freeze set in, coating roadways and sidewalks with a slick coating of ice. Soon it changed again to blinding, howling snow, crippling the elevated trains, horsecars, and cablecars all over town. Traffic in the streets was in a complete snarl; ferryboats running between New York and Brooklyn were stopped by the ice; and the single span between the cities, the Brooklyn Bridge, was perilous to cross in the gale-force winds. Snow was beginning to accumulate deeply in some parts of the city, covering the lower rungs of those fashionable brownstone stoops. On Monday morning many New Yorkers, unaccustomed to such blizzards and not realizing the severity of the storm, proceeded to their elevated stations as usual to get to their offices. Though the wind-driven drifts had risen to the tracks in places and snow touched the second story of town houses along some side streets, stalwart groups of businessmen and office workers made their way to the frigid, windswept el stations on that arctic, white Monday morning.

Col. Frank Hain, general manager of Manhattan Railway, tried valiantly to maintain service, but the storm proved overwhelming. Although some trains were operating in the first hours of the morning rush, most became stalled by mid-morning as snow and ice blocked their movement. A New York *Herald* reporter trapped over Sixth Avenue for four hours estimated that as many as 15,000 people were marooned in the snowbound trains that day, stuck high above the street in cold, wind-teetered coaches while a fierce winter hurricane swirled about them. The storm grew worse as the day progressed, and some authorities feared that the el trestle might collapse under the weight of the locomotives and the snow-laden cars. The fire department sent hook-and-ladder companies to the rescue, but they could not begin to reach all the passengers imprisoned on the slithery overhead rails. Some enterprising New Yorkers gathered ladders, propped them against the el, and sold safe passage to the street to el passengers at prices ranging from five cents to a dollar. The people in another stalled train unreachable by ladders had another idea; by lowering a cord to a saloon directly below them they were able to hoist up buckets of booze, and so they survived cheerfully and bibulously. Despite the hardships most urbanites sensed that they were part of a memorable event and braved the weather heroically. Not until Tuesday could the elevated resume any degree of business. Snowplows were dispatched in early morning to clear the track so that Col. Hain might send out some trains for the rush hour. Drawn by two locomotives rather than one, the trains crept along slowly at five or six miles an hour rather than the usual twenty-two. Even though there were numerous delays that morning because of the still icy tracks, the elevated had restored to the public at least partial rapid-transit service. The el was a good deal more fortunate than street transit, which continued to suffer for weeks from the mountains of accumulated snow, which disappeared ever so slowly.

It was not until the frantically busy traffic in the early twentieth century, just a few years after the system had been electrified, that the first real disaster struck the elevated. During the morning rush hour on September 11, 1905, a Ninth Avenue train headed downtown with motorman Paul Kelly at the controls was approaching the Fifty-third Street junction where the Sixth Avenue trains diverged to follow their own route and Ninth Avenue trains continued straight on south. Since the curve onto the Sixth Avenue branch was very sharp, train speeds were restricted to nine miles an hour. The destination of the trains was then indicated by an arrangement of disks on the forward car, and towermen lined up the switches for an oncoming train from these signals. Just ahead of Kelly's train a Sixth Avenue train turned into Fifty-third Street, but for some reason as Kelly approached the junction, towerman Cornelius Jackson failed to restore the track lineup for the Ninth Avenue route. Unexpectedly entering the sharp curve at thirty miles an hour (three times the safe speed set by the company) Kelly slammed on his brakes. While the first car remained on the tracks, the sudden halt in momentum threw the second car flying off the track, turning a complete loop in the air, and hurtling down into the street, sheering off its roof and coming to rest in a diagonal position, its front end on the ground and its rear caught on the trestle. The third car ran over the edge of the structure and hung suspended halfway out in space. The remaining cars, though derailed, continued on ahead down the Ninth Avenue line before coming to a stop. Thirteen passengers were killed, and forty-eight were seriously injured in the second car—the first car in the history of the New York Elevated to fall off the track.

Responsibility for the disaster rested between towerman Jackson and motorman Kelly; both denied any guilt, however, each blaming the other for the mistake. Jackson said the train had carried disks indicating a Sixth Avenue routing. Conductor J. W. Johnson, whose job it had been to set the disks, denied this, and the company backed him up, pointing out that there had been no confusion about the routing of the train, since station guards had correctly called out "Ninth Avenue Train" at every stop all the way down from Harlem. Some of the men would most certainly have noticed if the train had carried Sixth Avenue markers. But this was not the sole cause of the accident, because the track signals indicated the position of the switch five hundred feet before the train's approach. Regardless of the routing disks, motorman Kelly failed to observe that the switch was lined for the curve, probably because he was going much too fast to notice them. Since the IRT was at that time carrying something like 250 million passengers annually, and since this was the first occasion in the history of the elevated that a passenger within the doors of a car had been killed, their safety record was still not seriously discredited.

Despite the brief headway between elevated trains there were surprisingly few rear-end collisions. In this regard they were in sharp contrast with the experience of regular railroads, which suffered a plague of such wrecks. A number of rear crunches had taken place on the elevated over the years, but most were quite minor—except for one. The most serious collision in el history was a crash on the Ninth Avenue line on April 29, 1929, in which four persons were killed. The wreck took place near 167th Street in the Bronx on a section of the line jointly used by the Lexington Avenue subway and the el. Though the track was protected by automatic block signals, the stopping device failed to avert a serious collision because of the common practice of "keying" trains past the stop signals by holding down the tripper arm manually in order to allow trains to move up more closely during rush hours and thus increase track capacity. Just before the crash the subway motorman had stopped his heavy train of six steel cars at a red light just south of the 170th Street station, but since his train blocked a following train from pulling up to the platform and taking on its load, he was moved ahead by the "keyman," who said, "I'm going to key you by. Look out for that elevated train ahead." Although the motorman seemed quite rational and the visibility was perfectly clear, he put on full speed and ran blindly into the rear of the el train, shattering the last wooden car. Since he was one of the four people to die in the wreck, no explanation of his strange conduct was ever learned. After this crash there was naturally a great outcry against the practice of "keying by," and critics urged the procedure be banned, since it seriously eroded the protection of the automatic stop signals and created a substantial element of human error. Both elevated and subway officials defended "keying" as essential to avoiding excessive congestion and delay during rush hour, and thus the practice continued.

The accidents involving injury and loss of life that occurred on the New York elevated lines were, of course, regrettable no matter how infrequent. But since the total traffic must have approached thirty-five billion passengers during the eighty years of operation, it can be safely stated that riding the trains was far from risky and that the numbers of deaths were but an infinitesimal percentage of the total number of people they transported. For the elevated there were no tragedies of the magnitude of those that struck the subway—the BMT wreck at Malbone Street in 1918 that killed ninety-seven or the Times Square crash in 1928 that killed seventeen.

14 The Crush at Rush Hour

With the profound advances in urban transit engendered by the elevated roads it is no surprise that they were an overwhelming financial success. New Yorkers flocked to the aerial highways in such numbers, however, that the trains were soon positively surfeited with patrons. Morning and evening rush hours were pandemonium as hordes of darkly dressed commuters bound for their businesses pushed their way toward the cars—choking steep, narrow stairways and swamping change booths while the busy little trains arrived and departed, delivering and removing crowds, lifting clouds of dust, and jarring the sidewalks and the buildings and windows along the route. Station platforms were black with the struggling bodies, each person striving with all his energy to be first on the train when it arrived; yet often the throng who waited at the station rushed on board only to find standing room at best and sometimes hardly

that. Passengers leaving the trains at stations had to fight their way out of the cars. And since the stop was so brief many were carried one or two stops beyond their destination before they could manage to reach the car platform to step off.

So great was the population growth of New York that the elevated trains that in the 1870s had seemed to be a satisfactory solution to the city's transportation problems were no longer adequate to handle the traffic by the 1890s. As traffic increased steadily through the 1880s, the crush at rush hour grew worst on the Sixth and Third Avenue lines, those in the middle and closest to the core of the city. There was apparently no method of driving some of the persistent crowd to the other less busy trains on Second or Ninth avenues.

The chronic and painful overcrowding called forth some indignant protests from the press. The

The elevated was undoubtedly a crowded artery, but then so were the streets and sidewalks. Upstairs and down, Manhattan was pandemonium at rush hour. (From a German travel book, VON WUNDERLAND ZU WUNDERLAND, *1882.)*

Scientific American of 1896 complained that elevated stations were swamped with a struggling mob "in which the commonest laws of chivalry seem to be forgotten as strong men elbow frail women in the wild rush to secure the much coveted seat." The editors of the *Railroad Gazette* labeled the overcrowding a "special disgrace to our civilization," because they were particularly repulsed at the spectacle of ladies trying to hold onto straps far above their heads while a car swung heavily 'round a sharp curve. Usually quite rhapsodic about any commercial venture in the Empire City, guidebook editor James McCable (*New York by Sunlight and Gaslight*) complained that the dense crowds jammed into the cars during rush hour created an air so foul and malodorous that the danger of people contracting infectious diseases was very great (shades of the old horsecar stories). Equally nasty, he pointed out, were the bands of pickpockets, bullies, and ruffians who roamed station platforms in Irish neighborhoods, preying on the meek and unsuspecting.

William Howells's Altrurian Traveler was shocked at how, during rush hours, men and women were indecently crushed together on the elevated without regard for their personal dignity, the multitude overflowing from either end of the car onto the platforms. The el trains followed each other at close intervals, and at each station they made a stop of but a few seconds when those who wished to alight fought their way through the struggling mass. Those who wished to mount fought their way into the cars or onto the platforms, where the guard slammed an iron gate against their stomachs and in the faces of those arriving too late. Sometimes horrible accidents happened; a man clinging to the outside of the gate had the life crushed out of his body against the posts of the station as the train pulled out. With characteristic satire Howells concluded that in a country where people had such a dread of civic collectiveness of any kind, lest individuality suffer, the el patron was practically nothing in the regard of the corporate collectivities that abounded.

In a somewhat lighter vein, there was the story circulated by the *Evening Sun* about the coach of a celebrated football team who trained his men on the el. He would buy each man a dollar's worth of tickets and compel him to go up and down the Third Avenue road during rush hours. When an athlete objected to the harsh training, the coach said, "If you expect to win, you have got to go through the mill. Of course the disabled ones will be sent back, but the ones who come out all right will be able to rush through a brick wall." It goes without saying that after the Spartan-like training on the el they won their match, for at a critical moment in the field when their rush line wavered, the captain yelled out, "All aboard for the City Hall. Let 'em off first. Both gates. Hurry up!" The result was instantaneous; the line went through the opposition with such terrific force that the opposing team were scattered right and left over the field.

Despite some rather overdramatized reporting from the New York press corps, Manhattan Elevated provided about as good a service to the city as it was able—albeit crowded. There was, however, a definite need for expansion of their service to eliminate the crush, for the existing system was simply inadequate to handle the burdensome traffic. By 1890 the lines had reached the limit of their carrying capacity, and the number of trains could not be very much enlarged because they were already running as closely together as safety permitted. Faster and longer trains would be possible only if heavier locomotives were used, but heavier locomotives would require a stronger elevated structure, which would also mean a substantial increase in capital expenditures by the company for reconstruction. An alternative was to build additional elevated lines through the city or to expand those already in existence by additional third and fourth express tracks to far uptown and the suburbs. Again Manhattan Elevated was reluctant to engage in such expensive construction. Moreover, a great many residents strongly objected to the erection of any more elevated structures in public rights-of-way.

It was becoming clear to transportation engineers that steam-powered railroads had about reached their peak for urban rapid transit and that another solution would have to be found if cities such as New York would continue to grow and function. There had been a good deal of interest in cable railways in the 1880s, but such traction had so many disadvantages, both mechanical and operational, that it never proved satisfactory. There were also many supporters for underground railroads, but the principal problem with such travel was that until some satisfactory alternative for the steam locomotive could be found passengers faced the gruesome experience of riding in dreadful, dark tunnels full of nasty, sulphurous smoke and noxious fumes. The solution for both the elevated and subterranean roads was, of course, electric power; however, it would take years of experimenting and testing before a practicable electric motor could be developed for commercial use in rapid transit.

Before 1860 most of the experimentation with electric propulsion was made with batteries, though usually the operating efficiency was low and the work produced no immediate practical results. After 1860 attempts were made to produce electricity by some better method than batteries. In 1873 it was discovered that an electric motor could be made to generate electricity if rotated by mechanical means. At this time Dr. Werner Siemens successfully used electric power to drive a railway car. He built a short electric railway line that became one of the sensations of the Industrial Exposition in Berlin in 1879. Later he designed successful commercial transport lines in Lichter-

felde in Germany and between Portrush and Bushmills in Northern Ireland.

By the 1880s it was quite clear that electricity could be harnessed as a motive power in railways. During this decade several inventors in America set themselves to the task of working out the practical difficulties in the new technology. Thomas Edison became interested in the idea and built a speedy electric railway at Menlo Park, New Jersey, in 1880. Edward Bently and Walter Knight electrified the East Cleveland Street Railway in 1884 using a third rail in an underground conduit to supply current to the cars. This was the first successful third-rail line in Baltimore in 1885. Then in the next year Charles J. Van Depoele, a Belgian immigrant, was hired to electrify the entire fifteen-mile Montgomery, Alabama, street railway system; power was supplied by an overhead wire. It was then left to Frank J. Sprague to demonstrate the advantages of electric traction to the street-railway industry when he electrified the cars in Richmond, Virginia in 1888. Within two or three years there were hundreds of electric cars in operation in America.

In the face of mounting dissatisfaction over the grim, overcrowded conditions on his trains and from the growing interest in underground transportation, Manhattan Railway president George Gould (his father had died of tuberculosis in 1892) responded with a bagful of promises to improve his service. In a letter to the *Times* in 1896 he proposed an elaborate plan for expansion, including a new line along Tenth Avenue, a new crosstown line from Brooklyn Bridge to the West Side, and the laying of third tracks on the Second and Ninth Avenue routes for express trains. As might be expected Gould was highly critical of any proposed subway tunnels, praising instead the airy virtures of elevated travel. He could be quite voluable about future expansion and improved public services on his trains when any threat from the subways appeared, though the matter seldom went beyond empty rhetoric. Moreover, Gould usually ignored the suggestions of the Rapid Transit Commission, who had been encouraging el expansion, because he was much more interested in dividends than in the needs of the public. Consequently, el service on Manhattan Island remained for the most part unchanged, except for some third-tracking on Ninth Avenue, from the way it had been first laid out in the 1870s. Beyond

HARPER'S WEEKLY.

A JOURNAL OF CIVILIZATION

Vol. XXII.—No. 1132.]   NEW YORK, SATURDAY, SEPTEMBER 7, 1878.   [WITH A SUPPLEMENT. PRICE TEN CENTS.

Entered according to Act of Congress, in the Year 1878, by Harper & Brothers, in the Office of the Librarian of Congress, at Washington.

Through the gorge of Franklin Square shuttle the steaming elevated trains. One skirts the elaborate cast-iron facade of the Harper & Bros. publishers.

Ladies and boys wait meekly while a mob of men fight their way aboard an already packed platform on a rush-hour el train. COURTESY Harpers Weekly, FEBRUARY 8, 1890.

the island's boundaries, however, elevated rail service in the suburbs underwent considerable progress.

* * *

Until the mid 1880s the elevated railroads of New York City were all operated within the limits of Manhattan Island, commencing at South Ferry and extending northerly to termini on the East and West sides at the Harlem River. After the merger with the City of New York of the annexed districts in the lower Bronx, there was agitation for the extension of the rapid-transit lines across the Harlem River and into the new section of the city. Public meetings were held as early as 1875 and comprehensive plans discussed by the Rapid Transit Commissioners, but it was not until 1880 that the first company was formed to provide el service to that area. The Suburban Rapid Transit Company announced plans to construct el tracks in the Bronx through city blocks and over private property (rather than over the public streets as had been done in Manhattan), connecting them with the Third and Second Avenue lines of Manhattan Railway.

In the spring of 1886 the Suburban Company was ready for operation. They had built a steel drawbridge over the Harlem River to connect with the joint Suburban-Manhattan Railway el station at the head of Second Avenue and 127th Street and finished their tracks as far as 133rd Street, location of their only station. Within two years service was further extended to 170th Street. The Suburban Company's structures were at first built of iron and steel girders resting on granite and brick piers (except at intersecting streets, where steel and iron bents were built at the curb lines). Later on, standard elevated railroad structures were used similar to those in Manhattan. In 1886 the Harlem River branch of the Suburban line was opened, a single track spur to Willis Avenue and 132nd Street close by the New Haven Railroad Station. This branch served as a popular transfer for passengers between the Manhattan trains from the south and New Haven and Suburban trains from the north and east. In 1891, as a special operating feature, express trains were routed from the Wall Street area on the Second Avenue line to the Willis Avenue Station, so that passengers could transfer to New Haven trains for Morris Park Race Track in Westchester. The racetrack specials ran on all meet days until 1905 when all racetracks were put out of business after a state law was passed banning racing throughout New York.

Suburban Elevated Railroad was enjoying such good corporate health that its wealthy neighbor, Manhattan Elevated, began to hanker after them. As a result in 1891 Manhattan arranged a lease to operate all Suburban El properties in return for a yearly rental of $240,000. As soon as it took control, Manhattan Railway began to extend the structures from 170th Street along Third Avenue to 177th Street along with two new stations. Additional stations at 180th, 183rd, and Pelham Avenue (Fordham Road) were opened in 1901. From there tracks were laid through and over the grounds of Fordham College to the Botanical Gardens in Bronx Park. Soon after Manhattan Railway had taken control, it adopted a five-cent fare for a continuous ride from any point in Manhattan to the Bronx, so that a passenger could transfer without payment of any extra fare from all

The gala opening of the Brooklyn Elevated Railroad, May 23, 1885. COURTESY *Harpers Weekly.*

the East Side Manhattan Elevated trains to all those in the Bronx. The first through service of Second and Third Avenue trains from South Ferry to Bronx Park and from City Hall to the 177th Street Station (Tremont) were run in 1896. Through trains were then operated during rush hour period only; all trains except for locals were made through trains about two years later.

Elevated rapid transit in Brooklyn naturally followed about the same evolution as in Manhattan, though several more years elapsed before the growth of that city reached a point that required faster transportation than that offered by the horsecar. The Brooklyn Elevated Railroad Company was chartered in 1874 to run from the Brooklyn Bridge to Queens; but because of internal quarreling plus the unusual promotional difficulties, the project languished until 1884 when the company was reorganized and successfully financed. Finally in 1885 the first five miles of road

were placed in operation. In the meantime the Kings County Elevated was organized (largely by Boston capital) to operate the city's second elevated railroad. Their first section opened in 1888. Another pioneer line in Brooklyn was the Union Elevated Railroad, their first line running from the Long Island Railroad station through Flatbush and Hudson Avenues and connecting with the Brooklyn Elevated's line from Park Avenue to Fulton Ferry. In 1890 the Brooklyn Elevated and the Union Elevated were consolidated into a large system radiating all over Brooklyn City. In the mid-1890s all the various elevated companies were acquired by the Brooklyn Rapid Transit Company (BRT), which also absorbed most of the surface car companies of the city. It was the same BRT that joined the City of New York in the Dual System agreements of 1913 to organize the consolidated New York Municipal Railway Corporation. Though running similar locomotives and cars, the Brooklyn elevated railroad operations were a quite separate operation from those in Manhattan and the Bronx.

* * *

The great success of the elevated in New York prompted the construction of such railways in other cities in the world—Berlin, Kansas City, Chicago, Boston, Philadelphia, Liverpool, and Hamburg. However, the New York El, apart from being the first, remained the largest and most important elevated railroad system. By the time the last extensions were completed about 1920, there were just under 300 miles of single-track elevated railroad in operation in New York. With the exception of Chicago, other elevated railways were subordinate to some other transit system, such as an elevated-subway combination in Boston and Philadelphia or the elevated-trolley combination in Kansas City; whereas in New York the elevated was a unified network and the prime rapid-transit mover of people through four decades.

The city of Berlin built and operated a short viaduct elevated railway not long after New York's basic system was completed. The German capital's two-and-a-half-mile Metropolitan of 1882 joined two large railroad stations, the Frankfort and the Charlottenburg, with intermediate stops along the route. Being financed by the government, the $16,000,000 Berlin Elevated was very solidly and expensively constructed. Trains drawn by small steam-locomotives ran not over the streets on steel trestles as in New York, but through city blocks atop a masonry structure finished on the exterior like a well-built block of flats. At the street level the railway structure housed shops, restaurants, and offices. The stops along the route were all under lofty and spacious glass canopies, which were a picturesque feature of prewar Berlin.

In 1886 the rolling, hilly Western metropolis of Kansas City became the proud new owner of two miles of elevated railroad, running across the West Bottoms and connecting Kansas City, Missouri, and Kansas City, Kansas. The Inter-State Elevated, as it was called for a time, connected with a surface steam line at one end and a cable line at the other. The el was operated with small steam-locomotives until it went bankrupt in 1892, after which the corporation was reorganized, the service electrified, and the company absorbed by the Kansas City Street Railway. The steel structure, patterned after the ones in New York, continued in use until the end of the streetcar era, about 1950.

The Chicago Elevated, second largest system built anywhere, is still intact. It is now the only major downtown el network still in operation. The city's elevated railway operations began in 1892, when the first train ran over the high line of the Chicago and South Side Elevated. As in New York, trains were hauled by Forney-type steam locomotives. Within the next four years three other elevated railroad companies were established in other quarters of the city, rapidly spreading their tracks to cover the bulk of Chicago. Four separate roads were thus involved in shaping the basic el network, some intertwined financially and others independent. The four separate divisions—South Side, North Side, Metropolitan (West Side), and Lake Street—radiated out from the central downtown business district and extended to the northern and western suburbs. In the city center elevated trains circled the famous Loop in a constant parade, transporting passengers swifty above the confusion and congestion in the streets below. Chicago's elevated lines were somewhat earlier to electrify their operations than New York. In fact the Metropolitan West Side Company began service in 1895 with trains composed of electric motor cars and trailers. Their electric equipment was largely similar to that in use at the time on electric street-railways except that a third rail was used instead of an overhead wire to furnish current. Not to be outdone, the South Side Company took steps to electrify their line, resulting in their being the first to install Frank Sprague's system of multiple-unit control, the same system that was later adopted on the els in New York and eventually practically every rapid-transit subway or elevated in the world.

For a time the four lines operated as separate companies in Chicago with separate fares on each. Then in 1913 all of the roads were brought together and unified under a single system of operation to facilitate through routing between the north and south sides and to permit passengers to transfer to the west side lines without extra charge. The consolidation enable passengers to ride from Wilmette to Jackson Park, a distance of twenty-four miles, without changing trains. Since 1947 the city's elevated has been combined with the surface lines (streetcars, trolley buses, and motorbuses) to become the publically owned Chicago Transit Authority, thus unifying all local transit under a single management.

Next to New York's, the Chicago el system was the largest in the world, operating 82 route miles and 230 miles of single track, compared to New York with about 280 miles of single-track elevated. The Chicago el is still the backbone of the city's transportation network. Twice each day, in the morning and in the evening rush hours, elevated trains literally form a moving platform of cars that circle the downtown business district called the Loop. Though there are those people who are urging the city to tear down the old el as an eyesore, it is simply too useful and too practical to abolish, for without it, or some underground replacement, the daily flow of people in and out of Chicago would come to a halt.

In the same year that Chicago's elevated trains started running (in 1892), Liverpool, England, was borrowing the idea of overhead travel. The Liverpool Overhead Railway, or LOR, was the only elevated in Great Britian and the first el anywhere to be electricfied from the very start. Power was supplied by a central generating station and transmitted to motors on the cars; no separate locomotives were used. The steel superstructure closely resembled that of the Sixth Avenue el in New York. Built to serve the huge port area of the city,

*Some of the discomforts of elevated railroad travel as
interpreted by a newspaper cartoonist in 1888.* COUR-
TESY NEW YORK PUBLIC LIBRARY.

Mulberry Street station of the Inter-State Elevated, which connected Kansas City, Missouri, and Kansas City, Kansas. The structure was patterned after the New York El. COURTESY KANSAS CITY PUBLIC LIBRARY.

the six-mile-long Liverpool Elevated attracted thousands of visitors each year, because the high tracks offered a magnificent view of the docks and harbor.

There had been interest in building an elevated railway in Liverpool as early as 1877, when Anthony Lyster, son of the Engineer-in-Chief of the Mersey Docks, had traveled to New York to inspect the new elevated system under construction there. Though the Dock Board obtained the necessary Act of Parliament to build an elevated line worked by steam-locomotives, the years of delay put off construction until 1889, by which time electric power was available for traction. The LOR carried a heavy traffic of troops and workers during the grim days of World Wars I and II. Although heavily damaged during the Blitz bombing of 1940 and 1941, it was repaired and carried on valiantly. After the war years the line faced expensive replacement of most of its structure, and the directors voted to close it. Operations ceased in 1956, and the trestle was entirely demolished by 1959.

Boston opted for an elevated rather late in the game. In fact els were already going out of favor in some cities that owned them because they were thought ugly and disfigured the streets. Oddly, the Boston elevated arrived after the subway, which opened in 1897. *The New York Times* noted with irony that the nation's most conservative city should pioneer in building America's first subway. Since the underground was already established in

Boston, there was strong opposition to building an elevated. However, it was a preeminent fact that els were cheaper to build and so construction proceeded, and the first section was opened in 1901. Thus from the beginning Boston operated a combination elevated-subway system. The Tremont Street subway was at first a rather short, two-and-two-thirds-mile section in the city center used by trolleys. Elsewhere rapid-transit service was handled by electric elevated trains (the Boston el was from the beginning equipped with the Sprague system). Elevated trestles were never built in the heart of Boston City adjacent to the Common or the Public Gardens. At midtown el trains were underground through the central business district and then emerged on the other side and climbed back to their elevated tracks en route to the suburbs.

Following the opening of the first section of the Boston el between Charlestown across the Charles River to the north and Roxbury to the south, the system grew longer, with many additions and alterations to its original conformation. The scenic Atlantic Avenue spur meandered along the old Boston Harbor with its waterfront view of steamship berths and ferryboat slips. The spur was a colorful ride but was never a great success financially, though an especially bizarre misfortune attracted many visitors to the line. In 1919 a huge molasses tank belonging to the Purity Distilling Company on Commercial Street exploded, causing a two-million gallon wave of syrup to surge onto Atlantic Avenue and crash into a section of the el. When the line was abandoned in 1938, it was said that the smell of molasses still lingered there on hot summer days. In 1909 the el was pushed further south to Forest Hills, and in 1912 another elevated route was opened from the North Station to Lechmere Square in Cambridge. In 1919 service was also extended across the Mystic River to a terminal at Everett.

Today the elevated is still healthy in Boston, though plans are underway to eliminate some sections in favor of subway or buses. The northern section from Haymarket to Everett was scheduled for demolition in 1975, but the southern section from the city center to Forest Hills will probably be around for another decade, because public funds have been voted for a major reconstruction and repair of the line, along with the purchase of forty-five new cars.

The last of the elevated railways to be built in America was in Philadelphia. As in Boston, the people of Philadelphia were opposed to the idea of having elevated railroad tracks in the middle of their city. As a result they built a combination subway-elevated, which still runs underground through city center and is elevated in the suburbs. As originally built, the Market Street Subway-Elevated ran on an overground trestle from the western terminal at Upper Darby, over the Schuylkill River, then into a subway tunnel under Market Street and beneath the central business district to the east side of the city, there emerging again as an elevated serving the Delaware River docks and ferries to New Jersey. The terminal at Upper Darby is a junction between el trains and the suburban electric trains of the Red Arrow Lines and the Philadelphia & Western Railway. Construction of the project began in 1903; it opened in stages, and the entire system was in operation by October 4, 1908, using the Sprague system of electric train control, by then standard. Cost of the project was eighteen million dollars for the seven miles of subway-elevated and eighteen stations. The construction of the steel-trestle el structure was unusual by American standards in that it was built with rock ballast on a solid concrete floor to prevent any dripping into the street and to reduce train noise for people living and working nearby.

After the success of the Market Street line further rapid transit service was planned for Philadelphia. The Frankfort Elevated extensions was opened into northeast Philadelphia in 1922. From Second Street the Market Street trains emerged from the subway and became elevated, running six and one half miles to the end of the line at Bridge Street. Then in 1928 the Broad Street Subway was opened, running north and south and crossing the Market Street line at City Hall, where a network of underground passages and concourses connect to several underground stations, the Pennsylvania Railroad Station, and a number of offices and stores. Today the city-operated Market Street Subway-Elevated remains Philadelphia's most important transit artery, carrying more passengers than either the newer and longer Broad Street Subway or the surface commuter lines of the Pennsylvania or Reading railroads. The el is still largely as it was built, aside from the 1938 closure of the old Delaware Avenue portion

that served the river ferries. Also a new twenty-block subway tunnel under the Schuylkill and into West Philadelphia replaced a section of el structure in 1955. The remaining ten miles of vintage elevated railway in Philly are still as solid as the day they were built and give good service to the five million people who live there. Elevated travel is likely to stay around for many years in the Quaker City.

Combined el-subway systems appeared in Europe about the same time as in Philadelphia. Berlin's line, dating from about 1902, ran seven miles from Uhlandstrasse in the west to Warschauer Bruke just across the River Spree in what is now East Berlin. As built only one-and-a-half miles was tunnel, the remaining mileage running on a steel viaduct elevated. As the system grew between 1906 and 1930, however, it was built primarily as an underground, or U-Bahn as the Germans call it. Though severely damaged by the bombing and street fighting during World War II, the Berliner Hauptbahnen was patched up and is still running. Since the construction of the Berlin Wall ten years ago, however, it no longer crosses into the Eastern sector of the city. Hamburg, Germany's second largest city and the country's greatest port, also built an inner-city circular elevated-subway combination between 1906 and 1912. Of the total length of sixteen miles, about half was underground and half elevated railway running on a steel-and-stone viaduct. The el portion did not survive the wartime bombing and was not rebuilt.

A variant of the elevated railway, and in some ways comparable to it, is the monorail. A one-track railway with vehicles either suspended from or supported by a single rail track, a monorail is usually elevated and high enough to avoid colliding with street vehicles. Monorail travel has had some single-minded supporters over the years, though it has more often been regarded as a novelty. Most monorails have been associated with fairs and expositions, and at that most were either experimental demonstration lines or failures. Only two monorails have ever operated for any length of time as an everyday means of transportation: the Listowel line in Ireland and the Wuppertal Schwebebahn in Germany.

The nine-mile Listowel & Ballybunion Railroad, which opened in 1888, ran profitably for twenty-five years until it was closed because of highway competition; the Wuppertal Monorail is still running. The Schwebebahn, or swinging railway, is located in Wuppertal, a large, industrial city in West Germany, which sprawls eleven miles along the valley of the Wupper River. The line is the creation of Eugene Lange, who conceived of the idea of linking the towns of Barmen, Elberfeld, and Vohwinkel. Only instead of laying railroad tracks through the towns and paying great sums of money for right-of-way, he thought of suspending cars overhead and following for most of the route the narrow winding bed of the river. The nine-mile monorail (six miles over the river and three over streets) opened in 1904. Sloping lattice girders carry the two single rails forty feet over the surface of the water, sweeping in here and there to run down the center of a street before rejoining the river. Altogether eighteen elevated stations are located along the route, sandwiched between offices, department stores, factories, and apartments. The cars are suspended on a pair of twin-wheel tandem bogies, each bogie driven by an electric motor. Today trains run every two minutes during peak service, carrying a total load of four thousand passengers per hour in each direction. The ingenious Wuppertal overhead was damaged during the last German war but was rebuilt in 1946. Though considered a bit antiquated now, it is still a picturesque sight to see the cars wind through the river valley, hanging like fat spiders from their steel web.

After World War II there was renewed interest in monorail travel, especially in America because of its urban and suburban automobile congestion. Since 1950 several lines have been built, either as demonstrations or as novelties at expositions. Two demos were built at Cologne, Germany, by Dr. Axel Wenner-Gren, creator of the Alwed Monorail System. Two others were built in Texas in 1956, at Dallas and Houston, both for promotional purposes. The monorail erected for the 1962 Seattle Expo to carry passengers from the city center out to World Fairgrounds, a distance of 1.2 miles, proved so successful that it is still in daily use as part of the city's transport service. Curious about the possibilities of monorail travel, the Japanese built a short test line at the Tokyo Zoological Gardens. This was followed in 1964 by the permanent line linking the Tokyo International Airport with a city station of the Japanese National Railways, a distance of about eight miles. Pleasure monorails at the two Disneyland amusement parks in California and Florida also run on a daily basis, airily carrying visitors around the grounds.

15 The Spark of Genius: Sprague Electrifies the El

Electricity was moving rapidly into the life of New York in the 1880s, touching almost all quarters of the city in one form or another with its magical sounds and lights. Broadway was illuminated by electric arclights between Fourteenth and Twenty-sixth streets as early as 1880, and by 1885 telegraph, telephone, and lighting services were so common that their wires hung in heavy festoons above many streets and avenues. In 1887 the State Legislature gave a sort of grim sanction to the wondrous new energy when they passed a law decreeing that all criminals sentenced to death would be executed by electric current. As electricity became more commonplace as a source of illumination and communication, the idea of using it as motive power on the elevated railways came under discussion in the daily press, who were anxious to abandon the dirty steam locomotives that had become a plague to many neighborhoods of the city.

In envisioning electric traction as a cure for many of the city's transit ills, the press and city government were doubtless on the right track, but while the use of electric energy for lighting in homes and offices was in the throes of adolescence, the commercial application of electricity to urban transit was still in its infancy. It would take another generation of careful experimentation and testing before the science of electrical engineering would be able to furnish power for New York's elevated system.

Traditional railroad men that they were, most elevated railway officials were at best cool toward the idea of electric traction on their roads—all, that is, but one: Cyrus Field, whose earlier work of laying the transatlantic cable made him eager for innovation—especially electric innovation. With typical vision Field, at a meeting of the Manhattan Railway stockholders, spoke up for the then radical notion of electric traction, saying that

Junction of the Sixth and Ninth Avenue elevated lines, Battery Place Station, November 1938. COURTESY THE COLLECTION OF E. ALFRED SEIBEL.

nothing could be better fitted for the application of electricity as motive power than the elevated railroads: "The use of electricity in place of steam would bring this system of rapid transit as near to perfection as we are likely to attain in our day." Unfortunately his enthusiasm was not shared by his fellows, one of whom published abroad in the *Times* that Field was out of his mind in suggesting such a thing: ". . . a majority of us are determined to keep the elevated railroad system free of doubtful experiments. There won't be any electric motors on the elevated railroads. You can put that down in black and white." Indignant over public rejection by his colleagues in the company, Field decided, with characteristic bravado, that if no existing company would support him he would create his own organization to investigate the use of electric power on the el. He then formed the United States Electric Railway Company, drew up a contract for the use of a section of his New York Elevated tracks for testing, and established his nephew, Stephen Field, in a workshop in Stockbridge, Massachusetts, with a salary of two hundred dollars a month to come up with an electric locomotive that would revolutionize the New York El system. An electric engineer and inventor of modest reputation, Field had already developed a small, electrically propelled vehicle that had amazed the local Stockbridge townsfolk when he demonstrated the contraption on his farm in 1880.

The pioneer demonstration of an electrically powered train on the el was made not by Field, but by the English-born engineer Leo Daft, who in 1885 ran his locomotive the *Benjamin Franklin* pulling two cars over a section of the Ninth Avenue track between Fourteenth and Fiftieth streets. So that the experiment would not interfere with regular operations, Daft's engine was tried out at night when traffic had ceased for the day. The midsummer darkness created a dramatic backdrop for the locomotive, whose spectacular arcing lighted up the whole neighborhood in a bright blue glare. According to a *Times* reporter, who covered the event, when the engineer threw a lever to let the contact wheel down upon the third rail, a flash lit up the street like a big bolt of lightning. As the engine got underway, there was a whizzing sound as if an iron brush were revolving against a steel bar, and a galaxy of red and blue sparks flashed out from under the motor and fell into the street. Heading back down the track after turning around at Fiftieth Street, the electric train rumbled out of the station at a lively rate to the cheers of a crowd of people gathered on the sidewalk. The drive wheels of the motor were reflected in mammoth size on the fronts of the buildings along the route, and in nearby houses residents, startled at the fierce, blue glare dancing across their windows, thrust their heads out and stared like a flock of country folk at the circus. One bewildered barkeeper in a saloon ran out and chased the exotic-looking caravan up the street until he ran out of breath. Daft's train reached the lightning speed of twenty-five miles an hour, but since the company's Forney steamer could reach forty-seven pulling five cars, the Manhattan Railway management was not very much impressed by the *Benjamin Franklin.*

Later that same year another inventor conducted a similar series of tests with electric equipment on the elevated. He was Frank J. Sprague, a graduate of Annapolis, class of 1878, who had served in the Navy as an electrical engineer before joining Edison at Menlo Park, New Jersey. Since Edison was at that time stressing electric lighting to the exclusion of all other uses of electricity, Sprague left his employ after a brief sojourn to concentrate on electric traction, and in particular on the application of electric power to the New York elevated. In the winter of 1885, after making a thorough study of the el and its operational problems, he began making tests on the Thirty-fourth Street branch. By the middle of the next year Sprague was giving successful demonstrations of his electric car, which showed remarkable reliability, good pulling power, and fast acceleration; in fact the tests were so impressive that Edison seriously considered buying Sprague out. Even the staid el managers were themselves beginning to take notice.

So pleased was Sprague with his progress that he arranged for a special demonstration for the Manhattan Railway bigwigs—president Jay Gould, his lieutenants, and directors. Gould went uptown to the test site, and climbed into the lead car of the short train Sprague had rigged up for electric running, the other officials of the company trailing along behind the infamous financier at a respectful distance. Standing at the controls was young Sprague himself. The handsome young inventor now faced Jay Gould, maker and breaker of men and companies, ruler of thousands of miles of railroad, and master of New York rapid transit. This was the making of a Horatio Alger story; young Sprague would manfully seize the controls, turn the energizer in its sprocket, and the cars would shoot ahead smoothly, quietly, and smokelessly. Gould would be so impressed that he

This veteran steamcar built in 1880 was eletrified in 1902, when it became an instruction car used to train former engineers as motormen. Kerosene lamps on the front gates were needed in case of power failure. Third Avenue at 129th Street, 1955. COURTESY ROBERT M. VOGEL.

would immediately hire the brilliant inventor to convert his trains to electric operation. But, alas, it did not happen that way. When Frank Sprague stepped forward and turned his energizer, the train suddenly lurched forward, a fuse blew out in a blinding explosion, and Gould, terrified that he was about to be incinerated, tried to leap from the train. Fortunately the millionaire was restrained from jumping by some of his colleagues, who managed to calm him down after a few minutes, but when the car was finally brought to a safe stop again, he walked stiffly down the steps, shaking his head and swearing that he would never trust electricity again nor listen to the crazy man who talked him into trying it. The hopes of electricity on the elevated had been set back to the dark ages.

Yet another test of electric equipment on the el was made in 1887 by Stephen Field, who ran his locomotive on the Thirty-fourth Street spur pulling one car at eight miles an hour upgrade. El directors were unimpressed. Then Daft reappeared a year later with his rebuilt *Benjamin Franklin,* which did rather well hauling eight standard Manhattan Railway coaches, but again the el management failed to show interest. This ended the electric testing on the elevated for ten years, by which time the big changeover by street railways and elevated lines from steam to electricity had begun in earnest in Chicago.

Just why the officials of the New York Elevated were so uninterested in electricity remains unclear, for they must have been fully aware that their system under steam had just about reached the limit of its capacity. They were also aware of the objections of the people of New York to the noise and dirt that their locomotives created; in fact they had once commissioned Edison to make a study of the effects of the noise and vibration on the residents and structures along their roads, but there is no evidence that they implemented any of his suggestions. It appears that the elevated management simply ignored criticism, hoping that the public would come to overlook the discomforts of steam engines rumbling over their streets and that permission would eventually be granted for them to erect additional elevated structures, thus easing, for a time at least, some of the overcrowding. With these hopes in mind they refused to take any interest in electrification, even though it offered the potential for improving their service. Disillusioned because none of the directors of the el

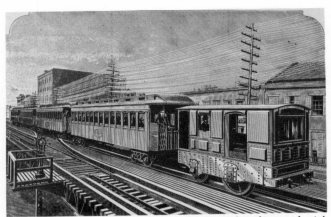

In the 1880s Leo Daft proposed to run the elevated with electric locomotives. This is his engine, the Benjamin Franklin, *which he hoped would replace the noisy steam dummies.* COURTESY Electrical Review, DECEMBER 1, 1888.

showed a serious interest in his work, Sprague took his ideas elsewhere to more appreciative audiences. His next project, to electrify a street railway in Richmond, Virginia, was far happier than his experience with the elevated; in fact it was so successful that it had a powerful influence on the future growth of electric elevated and street railways in the United States. From Richmond he went on to Boston, where he was awarded the contract to convert the largest street railway system in the country to electric running.

A significant development in electric railway history took place in 1895, when Sprague successfully demonstrated his new, multiple-unit operation by which a number of electric motors could be controlled from a single location. If installed on elevated railroad trains, every car would be equipped with electric motors, and each car could be operated alone or with any others. An entire train could be controlled from either end of any car, making for great flexibility of train makeup and permitting trains of any length. Heavy electric locomotives, such as those demonstrated by Daft and Field, which wasted power on short trains and damaged the elevated structure, would not be needed. With each car self-powered, the entire weight of the train including the passengers could be used as tractive power, and such individually powered cars would allow much faster acceleration. Moreover, the multiple-unit system on elevated railroads could provide extra capacity by permitting longer trains without expensive changes to the structure.

With the great advances promised from the new multiple-unit system, Chicago's South Side Elevated Company became immediately interested, and in 1897 Sprague signed a contract to electrify their entire system. The South Side installation worked remarkably well, from the beginning giving efficient and reliable service. Next to convert was Brooklyn, who began their changeover in 1898. That same year the London tubes banished steam locomotives in favor of electricity. With both the Chicago and Brooklyn elevateds taking the lead in electrifying their systems, the people of New York felt left behind, and they were growing annoyed with the management of George Gould, since his trains were still running on coal while those of other cities were running on faster and cleaner electric power. Reporters badgered George Gould constantly about the subject, though he displayed the same distrust of such motive power as had his father. In his message to the City Council Mayor Van Wyck, himself, spoke of the need of abandoning steam locomotives on the elevated so that "the people should no longer be subjected to the resulting nuisances of noise, smoke, and flying cinders."

Perhaps because of the public outcry, perhaps because of Chicago's increased earnings after electrification, Manhattan Railway directors began to reconsider their stand, for it was at this time during the 1890s that the elevated experienced its first competition from surface lines in the form of cable and electric street traction. The old, slow horsecar routes had given way to the cable first, then to electricity. Many of the old stages discontinued operation, as hundreds of new mechanically operated cars appeared on the streets, cutting heavily into the el's patronage and revenue. Also at this time was the beginning, however embryonic, of the automobile craze. Several wealthy men, including Alfred G. Vanderbilt and Harry Payne Whitney, had secured drivers' licenses from the police board to drive cars in the city. It was just the beginning.

Under the circumstances the directorate of the elevated began seriously to weigh the benefits of electrifying their trains. Though conversion would be frightfully expensive (as much as eighteen million dollars) the company would be able to sell off 225 locomotives, which were consuming 226,924 tons of coal and 535 million gallons of water a year: the savings in fuel alone would be enormous.

And conversion would mean an immediate solution to their overcrowding, because with electricity they could start running longer trains. That electricity would make New York a cleaner, more pleasant place to live in for millions of people was probably not a consideration by the company.

When Willaim E. Baker was named general superintendent of Manhattan Railway in March 1899, his appointment seemed a clear indication that the system was ready to go electric, for Baker, an electrical engineer, had helped to electrify the West End Railway in Boston and the Metropolitan Elevated in Chicago. Shortly thereafter, Edwin Gould announced that his company would opt for electricity, and within a few months the actual changeover began. Work first began on the huge power station at the foot of 74th Street and the East River. The 100,000-horsepower generating machinery manufactured by the E. P. Allis Company of Milwaukee was at the time the largest such plant in the country. The first test of the new trains was held on the Second Avenue line in the early winter of 1900 at the same time that Lieutenant Winston Churchill, MP, was visiting in New York City to lecture on the Boer War. The test being successful, the company proceeded with the changeover of the entire system. The General

Electric Company of Schenectady, New York, was awarded the contract to furnish all the motors, controllers, and other equipment for the cars at a cost of three million dollars, though the actual installation was done at the Manhattan Railway shops. The initial order for new coaches was placed in June 1901, fifty to be built at the former Jackson & Sharp Delaware Car plant in Wilmington, Delaware (which had been acquired by American Car and Foundry), and fifty by the Wason Manufacturing Company of Springfield, Massachusetts, builder of many earlier elevated cars. Wason also furnished the millwork for the motormen's booth to be installed in five hundred old steam coaches.

While steam locomotives hauled trains in regular service, electric test runs were sandwiched in between for trying out the new equipment and for training locomotive engineers in the operation of electric cars. By January 9, 1902, all was ready for the official start of multiple-unit train operation on the Second Avenue line. To inaugurate the splendid, new service, the company invited two hundred guests to ride a special six-car train over the entire Second Avenue Division. Beside George and Edwin Gould were such titans as John D. Rockefeller, Jr.; H. H. Vreeland, president of the Metropolitan Street Railway; John McDonald, the subway contractor; and A. N. Watts, superintendent of the New York Central Railway. As the train pulled out of the South Ferry Station on that

Built in 1890, this wooden car was electrified in 1902. It ran in service until 1942 when it was sent to California for industrial wartime use. COURTESY THE EDWARD B. WATSON COLLECTION.

New York in 1910 was an electric city. The Sixth Avenue elevated stretches along at the east of Herald Square while on the left runs Broadway, with its glittering streetlights, electric billboards, and electric automobiles, trucks, and trolleys. COURTESY THE SMITHSONIAN INSTITUTION.

Friday morning, it sounded a long blast of its compressed-air horn, whose strident note—so far different from the familiar mellow throatiness of the steam whistle—heralded a new era in New York transport. The strange sound caused a flurry of excitement from people in the street and in nearby buildings who stared curiously at the peculiar, engineless caravan. The new train made a dazzling appearance that dark January morning, since all the incandescent lamps were lighted—twenty-three bright bulbs in each coach, not in clusters here and there, but running in a line through the length of the ceiling. In their interior arrangement the cars were similar to those used during steam years, except for the motormen's cabs in the power cars.

Full electric service on the Second Avenue route began in March, on the Third Avenue line in August, and on the Sixth Avenue in November. The oldest elevated line in the city, the Greenwich Street-Ninth Avenue line, was the last to change over, in spring 1903. Within but a brief time the new Spraguian multiple-unit trains had usurped the authority of the gamey little Forney steamers, who were all ultimately banished for menial labor in the far corners of the globe. The end of the steam era on the el had come.

The benefits of electrification became apparent to the company almost immediately: train acceleration was up twenty to thirty percent over steam. From a full stop the new trains could hit fifteen miles an hour in as many seconds or about twice as fast as the old steam-hauled cars, meaning quicker movement of trains on all lines. Also, by eliminating the locomotive, one more car could be added to a train (which meant running six rather than five cars maximum), substantially increasing the capacity of the system at rush hour. There were other happy side benefits of electrification: electric lighting and heating, new block signals that automatically halted a train if the motorman ignored a red light, and remote-controlled doors and cab signals allowing all the doors of a train to be regulated by one man and permitting the train to start instantly when all the doors were closed. After electrification it was also fairly easy for the company to electrify its stations for lighting, install electric elevators at especially high sections of track (such as at 110th Street), and add fast escalators at the bustling midtown stations (such as Herald Square).

*The el tracks on Sixth Avenue and Thirty-fourth Street.
By the time the el was electrified in 1903, New York had
probably become the most brilliantly lighted city in the
world. The night sky was luminous from the reflection of
scores of bright electric signs and thousands of office
and store windows.* COURTESY PRINTS AND PHOTOS
DIVISION, LIBRARY OF CONGRESS.

Some patrons compained of rough, jerky starting and stopping of trains after the changeover, but
in general New Yorkers were delighted with the
new, clean, and relatively silent power. Yet it is
doubtful whether the elevated's officials were
really very much moved by all the years of
complaint about smoke, dirt, cinders, and noise
from their trains. Surely, the electrification had
not occurred because the company wanted to add
to the comfort of city residents by dispensing with
the steam locomotive; rather the changeover to
Sprague's multiple-unit system came about because of one paramount virtue—it reduced costs
and increased revenues markedly. For fiscal year
1904, the year following the conversion, expenses
for the company were down $99,000.

Within but a few years New York had become
an electric city, probably the most brilliantly
lighted metropolis in the world. The night sky over
Manhattan had become positively luminous from
the reflection of scores of bright electric signs,
thousands of brilliant office and store windows,
and the flashing of countless elevated trains that
scurried like brisk yellow bugs along the brick
canyons of the avenues. Broadway, especially,
was scintillating in its electric glamour—a gaudy,
glittery trail, a "Great White Way."

16 The Enemy Below: Competition Threatens the Elevated

Unable to extort reform of the painfully over-crowded elevated railroads from Gould and his Manhattan Railway directors, many enlightened New Yorkers had turned to other measures to improve city transit. Since private enterprise appeared so obdurate to the needs of the public, a group of citizens began advocating public ownership of mass-transit lines as a possible solution. A rally was held at Cooper Union in support of a bill permitting the city to operate its own rapid-transit service, and although the bill failed to pass in the legislature, reform was clearly in the air. One of the men to speak up for public ownership at that meeting was Abram S. Hewitt (son-in-law of Peter Cooper, who had backed Charles Harvey's first elevated on Greenwich Street), who became mayor of New York in 1887. Hewitt, whom many regard as the father of modern rapid transit, advocated use of city credit for construction of

additional transit lines, either elevated or under-ground, the operation of which would be handled by private corporations.

The machinery for the eventual creation of a comprehensive transit plan for New York came about in 1891, when the State Legislature finally permitted the city to name its own rapid-transit commission, lay out routes, adopt plans for new routes, be they elevated or subway, and put the franchise to build and operate them up for public auction. Under the new law William Steinway became chairman of the commission with William E. Worthen as chief engineer and William Barclay Parsons deputy chief engineer. Within five months of its creation the new commission decided to build an underground railroad and announced a tentative route running up Broadway. After the approval of the mayor, bids were solicited for the actual construction; but since no bids appeared,

The el's third tracking and "express hump" construction undertaken from 1913 to 1920 resulted in a jungle of ironwork in some parts of lower New York. This photograph was taken in 1955 at the time of the closing of the Third Avenue line. COURTESY ROBERT M. VOGEL.

the project failed for want of capital. It was widely believed that opposition by the Gould syndicate was responsible for the failure to sell the obviously valuable subway franchise. Gould was naturally loath to the thought of a competitive transit system in the city and was powerful enough financially and politically to present a considerable obstacle to it. Having failed to enlist capital for their subway, the commission then appealed to Gould to build extensions to his elevated lines, but so secure was he in his transit monopoly now that the subway project had aborted and that he ignored the appeal. As a result of their failure to build either the subway or bring about expansion of the existing elevated lines, the Steinway Commission was judged a failure by the Legislature in Albany, who abolished the body and created a new one, the Rapid Transit Commission of 1894, of which Alexander E. Orr was made chairman with Wil-

liam Barclay Parsons again as chief engineer. Early in their deliberations the new board, like its predecessor, reached the conclusion that the only way of meeting the difficult transit situation was to build underground railroads, and the people of New York concurred wholeheartedly, because they were sorely wearied with the inadequate service supplied by the elevated and of the presence of their unsightly structures in the streets of the city.

The new commission had the support of the mayor, a number of prominent businessmen, and the Chamber of Commerce. Led by Abram Hewitt, a campaign was waged to enlist support for public ownership and private operation of the projected subway. Ths question of the ultimate ownership was decided on November 6, 1894, when a referendum was held. An overwhelming majority of voters chose public ownership of the new roads, thus making possible the use of the city's credit in providing the millions of dollars needed to bring the great project to completion. Having received the mandate of the people, the

Construction photograph of the new elevated in Queens, taken in 1916. This was an extension of the Second Avenue el over the Queensboro Bridge.

board proceeded to push ahead with the difficult work of selecting the final routes and perfecting the engineering details of the plan.

The idea of a publicly owned subway competing with their heretofore unmolested elevated roads was most unsettling to the Gould interests, for though the patriarchal Jay Gould had died in 1892, Manhattan Railway was still controlled by the family. Their first reaction was to head off the subway by offering to build extensions to their routes in the suburbs. The Rapid Transit Commission, however, refused to give their needed sanction unless the el also expanded their service in lower Manhattan, where the facilities were so badly clogged. Moreover, the el management had made no offer to the city of rental for use of the public streets for the new lines nor pledged any time limit for completion of the work. The elevated's diversionary tactic to stave off the building

of the subway was thus unsuccessful. Neither the power nor the wealth of the elevated magnates could delay the coming of the inevitable—competition.

Principal financier of the subway project August Belmont, who had earlier been a backer of elevated railroads in Brooklyn, recognized that the elevateds were obsolescent, saying, " . . . I realized that the elevated roads, while they had served a most useful purpose, must be superseded or supplemented by fast electric underground roads." Despite wild skepticism and profound problems with the financing and organizing of the subway project, work proceeded rapidly under the direction of contractor John B. McDonald. Ground was broken in spring 1900 at City Hall, and the first section was opened to the public on October 24, 1904, from City Hall to 145th Street. In an attempt to coordinate elevated and subway service in the best way possible, the Interborough Rapid Transit Company (IRT), which had been established to operate the subway, leased all the

160

Manhattan lines, paying as rent seven percent of the capital stock, $60,000,000. This placed IRT in control of practically all rapid-transit lines in Manhattan and the Bronx.

The traveling public of New York took kindly to the new subway line, and within a few days after the official opening the tube began to show signs of crowding during rush hours. Before the first year was out, complaints of inadequate service were arriving at the Rapid Transit Commission. As the traffic grew, so did complaints of inadequate service increase in such numbers that by 1906 the commission was forced to appoint a subcommission to investigate them. The crush at the City Hall-Brooklyn Bridge subway station was particularly notorious, as there a flood of incoming commuters over the bridge from all parts of Brooklyn and Queens were discharged to find their way to their places of business in Manhattan. Denunciation reached such a peak of 1906 that a group of prominent businesses invited Gov. Charles E. Hughes to witness for himself the horrors. The governor saw and was apparently shocked, for he set to work to devise a new agency to bring about the needed relief.

New York's public transit problems had thus come full circle, going from the horrors of the horsecar to the squeeze of the elevated trains, and finally the crush on the subway. Ironically, the new subway was such a success that people clamored for more subterranean roads to be built all over the city, but the progress in construction was rather slow, since such pioneer labors always

Looking north from the Ninth Street station on the Third Avenue el as a train leaves on the local track. The photograph was taken from the upper express platform, September 1942. COURTESY THE PRINTS AND PHOTOS DIVISION, LIBRARY OF CONGRESS.

take more time than those undertaken after the pathway has been cleared. In the wake of discontent over the delay in subway construction the Rapid Transit Commission fell, for instead of praise for creating the subway they were criticized for not expanding it fast enough. Governor Hughes next created a Public Service Commission to supervise the elevated and underground railroads in the public interest and to devise a comprehensive plan for rapid transit through the five boroughs of the city. In 1897 a new charter for Greater New York had come into existence uniting into one municipality Brooklyn, Queens, Richmond (Staten Island), the Bronx, and Manhattan. It was the Public Service Commission that eventually made possible large-scale expansion of the city's elevated system between 1913 and 1917, launching the most notable improvements in el service since 1878.

Until greater New York could afford to complete its ambitious and staggeringly expensive subway system into all areas, the city was obliged not only to keep all the existing elevated routes but also to construct new ones, since they were cheaper and quicker to build than subways. Elevated railroads may have seemed a rather dated

Elevated construction about 1913 to 1916 doubled the tracks at Chatham Square. Since there was usually not room for the company to widen their stations in lower Manhattan to provide additional service, they simply built the center express track above the two lower local tracks and erected new platforms and stairways. COURTESY THE COLLECTION OF E. ALFRED SEIBEL.

Until greater New York could afford to complete its expensive subway system, the city was obliged to expand its elevated routes since they were cheaper and quicker to build than subways. This is part of the Queensboro Bridge el along Jamaica Avenue in Queens. C. 1917.

and second-best solution to the city planners of the new century, but they nevertheless continued to endure in vigorous good health for another generation after the newer, more glamorous competition had enter the field.

The combined subway and elevated divisions of IRT were carrying over 1,500,000 passengers daily by 1907, so that it was important for both services to expand as soon as possible. This urgency eventually led to what were called the Dual System conferences, which worked out the complicated arrangements of bringing about the construction and coordination of the over- and underground rail expansion. Plans called for doubling existing rapid-transit service by providing for new subway elevated lines, expansion of existing elevated lines, and creation of several elevated-subway connections in Manhattan, Bronx, Brooklyn, and Queens. It was called the Dual System officially because two companies, the IRT and the BRT (Brooklyn Rapid Transit) would be linked together under city auspices. It was also a dual operation of elevated and subway lines, operated by the two companies with the city retaining the right to supervise them and share equally in the profits.

The four major extensions to the elevated system were the Webster Avenue Line, the Eighth Avenue and 162nd Street Connection, the Queensboro Bridge Line, and the West Farms Subway Connection. The Webster Avenue Line was an extension of the Third Avenue elevated from its terminus at Fordham northward through Webster Avenue to Gun Hill Road and eastward to a junction with the elevated extension of the Lenox Avenue (Sixth Avenue) branch of the subway on White Plains Road. The Eighth Avenue and 162nd Street connection was an extension of the Sixth and Ninth Avenue el line from their terminus at 155th Street and Eighth Avenue over the Putnam Division Bridge over the Harlem River and through East 162nd Street to a junction with the Jerome Avenue elevated extension in River Avenue of the Lexington Avenue subway. The Queensboro Bridge elevated was an extension of the Second Avenue line over the Queensboro Bridge to a junction with the new rapid transit lines to Astoria and Corona; another extension took the Broadway elevated line out Jamaica Avenue to Grand Street in Queens; the Liberty Avenue line was an extension of the Fulton Street elevated line from the Brooklyn-Queen boundary out Liberty Avenue to Lefferts Avenue, Richmond Hill in Queens. The West Farms subway connection was an elevated line joining the Third Avenue el with the Lenox Avenue branch of the first subway and running from about 143rd Street through Willis and Bergen Avenues to the subway line, which was an elevated road at that point. Construction work on the enormous project began in 1913 and was completed in various stages between 1917 and 1920.

Before the signing of the Dual System contracts the old tracks of Manhattan Elevated were badly in need of remodeling and enlarging to handle rush-hour express trains. As a result the Public Service Commissioners also undertook an ambitious reconstruction and enlarging project at the same time as the extensions to increase el service. This work generally consisted of adding third and sometimes fourth express tracks to supplement the old double tracks and the construction of express stations to the Second, Third, and Ninth Avenue elevated lines. The work, usually known as third-tracking, called not only for the laying of additional track between the existing ones, but also for the strengthening and rebuilding of countless foundations and superstructures to support the additional weight. Many stations had to be rebuilt, often by adding express humps, a term used to describe the double-decking that allowed passengers to change from locals to express trains.

A new elevated railroad trestle marches past an old country villa. Demands for expanded public transit from 1912 brought about construction of new el lines in parts of Long Island where but a few years before had lingered a rural atmosphere.

As part of the new Dual System Project, the Interborough Company began work on March 13, 1914. Construction was finished by January 1, 1916, at a cost of twenty million dollars. For twenty-one months the aerial work of building the new lines and rebuilding the old ones went on without suspension of traffic either on the railroad or below in the streets. The entire job was completed aloft, often in congested, narrow spaces not more than twenty feet wide. Altogether fifteen miles of trestle and track were built, nine miles of old line rebuilt, 638 foundations reconstructed, twenty-nine stations remodeled and enlarged, and seventeen stations with double-deck express platforms built. The old Harlem River bridge was replaced with a new six-track double-level span to carry trains more expeditiously to the Bronx. From Chatham Square to City Hall two new tracks

were superimposed on top of the old local tracks. Pillars and structure for the double burden, therefore, had to be twice as strong. While the new columns were set in their concrete foundations, traffic was borne on a temporary wooden structure, heavily braced to withstand the weight and vibration. In many places, particularly along the Bowery stretch, engineers found it necessary to excavate far below the surface of the street to find proper foundations for the columns.

The most difficult feat of construction was at City Hall. There street congestion created by the entrance to the Brooklyn Bridge was greater than anywhere else in Manhattan. During rush hours elevated trains ran forty seconds apart. Some nine hundred trains a day carrying 120,000 passengers left the terminal, while at Chatham Square, where South Ferry trains crossed the tracks from City Hall, 1,892 trains passed the station every twenty-four hours. With such a torrent of traffic going by, the old double-track Park Row line was increased to four tracks on two

An express hump station looms over the old local station at 125th Street on the Third Avenue el. The appearance was the same at most express stations on the Second, Third, and Ninth Avenue els. COURTESY ROBERT M. VOGEL.

Second Avenue el, 1929, etching by Reginald Marsh. COURTESY PRINTS AND PHOTOS DIVISION, LIBRARY OF CONGRESS.

levels. At City Hall a double-deck station with six platforms replaced the old terminal with three platforms. At Chatham Square engineers had to separate two streams of train—one going to Brooklyn Bridge and the other on to South Ferry. Without interrupting the flow they eliminated the grade crossing and replaced it with a double level of tracks. A new station was then built serving both levels with eight tracks. Consequently, a Second Avenue el train for City Hall took the upper level tracks at Chatham Square after passing under South Ferry trains on the Third Avenue line.

Since there was usually not sufficient room for the company to widen its stations to provide additional service, it simply built the center express track above the two lower tracks and erected new platforms and stairways. Eleven of these hump-style express stops were built on Manhattan during the years of reconstruction and six others in the Bronx. These express humps or undulations, which became a special feature of the new elevated service in the 1920s in New York, were first proposed when the el was built in 1872. Engineers even then recognized the usefulness of such humps in retarding incoming trains and accelerating departing ones. But they were overruled because of the additional stairs passengers would be forced to climb. They were adopted in 1914 partly because of the gain in train speed but largely because the narrowness of the streets would not permit island platforms.

While elevated railroads are often considered an essentially nineteenth-century phenomenon, the fact remains that New York's el was not only healthy and thriving in the twentieth century but grew notably between 1913 and 1920. Not only were forty miles of new track added to the system in the boroughs of Bronx and Brooklyn, but in Manhattan service was greatly expanded despite competition from the new subway trains. The el construction boosted the number of trains serving New Yorkers in the evening rush hour, for instance, from sixty-seven express trains with the old facilities to 131 with the new. Although el patronage had been pretty much stationary for many years, the first year's gain after the opening of the third tracking was thirteen percent.

Although the elevated has vanished in Manhattan, New Yorkers in other boroughs continue to face the presence of the el in their neighborhoods. Of a total of 230 route miles of the city transit rail system, about seventy-one miles are still elevated in the Bronx, Queens, and Brooklyn. According to the city Environmental Protection Administration more than 593,000 New Yorkers live within 150 yards of elevated or grade-level subway tracks. In fall 1977 a hundred people from various civic groups gathered in Brooklyn to protest the el through a new organization called the "Big Screechers: People Screeching Against Elevated Train Noise." The el undoubtedly continues to bring blight though it also remains an efficient, though perhaps antiquated, transportation system. To replace it with underground would be expensive—about twelve billion dollars.

17 Let the Sun Shine In

Wearing safety goggles, an orange hard hat, and asbestos gloves, Manhattan Borough President Hulan Jack stood on a hoist truck on the corner of Third Avenue and Forty-second Street weilding a wrecker's torch. He cut a steel girder from its two upright supports; then a crane lifted the crossbeam and the uprooted columns onto a flatbed truck that hauled them away, as a crowd of onlookers cheered and television and movie cameramen recorded the scene for history. So it was in the early afternoon of February 16, 1956, that the end came to the last of New York's famous old elevated railroads that had straddled four of Manhattan's principal avenues since the 1870s.

As it ran in the shadows of new steel and glass skyscrapers, the el was a link with the city's past. While it endured there were memories of diminutive steam dummies whose puffing chimneys darkened wash lines and Victorian waiting rooms above the street with blue sunlit stained-glass windows and black potbellied coal stoves that cast a glow of warmth to travelers on cold days. Many people still recall the vivid imagery of the el—the gabled pavilion roofs and cupolas of the stations that swayed slightly and rattled noticeably when a train passed, the approach of a train from the platform, a weak headlight, the colored marker lights, the ding of the gateman's bell, the click of the door engines closing, or the whine of the electric motor cars as they struggled to accelerate. New Yorkers of the future will not experience the vicarious thrill of meandering down through the man-made canyons of lower New York on an el train or the twisting rails through Pearl Street and the old dock section near Coenties Slip curve, the sleeping derelicts on the Bowery sidewalk, or the breathtaking view from the Great Serpentine Curve at 110th Street and Columbus Avenue.

As it endured into the era of steel and glass uniformity, the ornamental Victorian station architecture was a nostalgic link with the city's past. Ths depots were meant to be looked at and to give pleasure in being seen. Photograph of the Third Avenue station at Seventy-sixth Street, 8:30 A.M., September 1942.
PHOTO BY MARJORY COLLINS, COURTESY PRINTS AND PHOTOS DIVISION, LIBRARY OF CONGRESS.

At the time of its death in 1956 most people no doubt thought of the el as hopelessly antique, ugly, and rather outlandish. But it was not always so. Seventy-five years earlier the el was new, quite grand, and wonderful—the last word in transportation. Cities that had them boasted of their modernity. Passengers were invited to leave the world of the streets and ascend to a higher plane, from which the old sidewalks and buildings took on a new perspective. The great success of the els in New York prompted the construction of elevated railways in other cities: Boston, Philadelphia, Kansas City, and Chicago in this country and abroad in Berlin and Liverpool.

A ride on the el was for many years an adventure, but eventually it ceased to be a fashionable mode of travel, though it remained long after as a fast and enjoyable means of getting around. In time the el lost its popular support except for a few sophisticated patrons and sentimental devotees. For seventy-seven years the el had been part of the New York landscape, but beginning in the 1930s it began to come down—the Sixth Avenue line in

1938, the Ninth in 1940, the Second in 1942, and finally the Third Avenue line in 1955. In the 1960s most el operations were also discontiuned in the other boroughs of the city. Brooklyn's old Myrtle Avenue line came down in 1969, and in 1973 the eighty-seven-year-old Third Avenue elevated in the Bronx was shut down as part of Mayor Lindsay's program to modernize transit services. A fleet of diesel buses replaced the antique section.

Manhattan's elevated railways had of course been doomed for nearly half a century, ever since the first subways was completed in 1904 and electricity replaced steam. Their eventual demise was only a matter of time as the city laid plans to extend its subway network to all the boroughs of New York. Subways were thought to be faster, more efficient than the els, though no cleaner nor more comfortable. Despite the notion that the subway was quicker, the speed of the el was substantially the same as that of the subway. The el's advantage lay in its rambling trajectory replete with images of New York City, which the subway (except for brief aerial excursions) lacks.

Though the el's patronage remained healthy through the 1920s and early 1930s, such travel grew unpopular. Most New Yorkers considered it dreadfully old-fashioned, a nasty eyesore. One city father publicly labeled it "a blight to the borough" and another called it "a clattering iron serpent that sprawls down the length of our island." Because of its appearance there was great pressure put on the city by property owners and civic organizations to tear the el down in the belief that real estate would benefit. During these years of decline the el was neglected and it took on a seedy, unkempt look. The old, gingerbread stations were in bad shape and the waiting rooms grew dingy. The last of the el expansions having taken place in 1920, the system became static, then moribund. Service was cut back, and several spurs were closed, the Fifty-eighth Street branch in 1924 and the Thirty-fourth Street branch in 1930.

The depression hit the els quite hard, and the beginning of the end came when the IRT and the Manhattan Railway Company went into receivership in 1932. This adversity soon led to the demolition of the Sixth Avenue elevated. Since the company could not pay its taxes, the city stepped in and condemned the property. In mid-1938 the court authorized the sale of the line to the city for

Underneath the latticework of the trestle was like an arcade, not really such a bad place. And the sun filtered through the bower creating a pattern of light and shadow that enchanted many artists. Greenwich and Liberty streets, c. 1935.

The sweep through Battery Park, 1935. COURTESY THE COLLECTION OF E. ALFRED SEIBEL.

the sum of $12,500,000, of which $9,000,000 was to be deducted for delinquent taxes outstanding against the entire Manhattan Railway system. Service was discontinued on December 4 and replaced by the new city-owned Independent Subway, built beneath Sixth Avenue.

Bids for the sale of the structure and stations had been advertised several days before actual possession by the city. It was sold to the Harris Structural Steel Company of New York, whose men began almost at once to raze the gaunt trestlework. On December 20 Mayor Fierello La Guardia held a brief ceremony at the corner of Sixth Avenue and Fifty-third Street, wielding the first acetylene torch in the process of cutting down the ironwork.

The actual demolition proceeded rapidly—one span at a time and with little disturbance to traffic in the streets. A crane lowered rails, ties, signals, etc. to the street, where they were hauled away in

Creating an abstract tracery against the September sky—the Second Avenue elevated in the midst of demolition. Looking south toward the stumps from Thirteenth Street, 1942. PHOTO BY MARJORY COLLINS, COURTESY PRINTS AND PHOTOS DIVISION, LIBRARY OF CONGRESS.

big flatbed trucks. Meanwhile the rivets securing the girders to the columns were burned out with torches, and after lashings were made fast to the beams a crane tugged and unseated the entire span, which was then lowered to the street and cut up. Removal of the spans proved and easy job, but the demolition of the stations along the line took more time. By April 1939 the last section had vanished, most of the steel having been sold to Japan. With the el pillars no longer impeding the flow of traffic nor the superstructure bringing perpetual twilight to the street, there is little doubt Sixth Avenue looked wider and airier than anyone could ever remember.

The razing of the Sixth Avenue line foreshadowed the eventual demise of the entire el system of Manhattan in addition to many lines in the Bronx and Brooklyn. When the city acquired the IRT and the BMT in 1940, plans were made to abandon all el operations as soon as possible, substituting publicly owned subways and motorbuses run by the New York City Board of Transportation. Such undoubtedly is the nature of progress.

Consequently, the Ninth Avenue elevated dis-

continued operations in 1940, and on the East Side the Second Avenue line closed down in 1942. After the second German war, agitation was renewed to rid Manhattan of its last remaining elevated railway. In January 1948 Major O'Dwyer's planning commissioners recommended the building of a subway under Second Avenue and demolition of the Third Avenue el, "last remnant," they said, "of an engineering monstrosity." And so it was in such an atmosphere of disfavor that elevated railway service on Manhattan Island came to a close. On the curiously ominous day of Friday, May 13, 1955, the last train ran on Third Avenue, jammed to overflowing with nostalgiacs, souvenir hunters, newsmen, and cameramen. The sentimental journey started from Chatham Square near Chinatown at 6 P.M. and ran to the end of the line at the Bronx 149th Street station. Along the Bowery groups of drinkers emerged from taverns and bars to look. At Forty-second Street there were crowds in the street to watch the last train rumble by, and all along Third Avenue the upper windows were filled with people witnessing the passing.

On August 2, just a day before the demolition was to begin on the Third Avenue line by the Lipsett Brothers, the East Side bid a final farewell to the el. In the sizzling, late-summer heat a cavalcade of ancient automobiles carried city representatives under the el from Chatham Square to 125th Street, stopping every few blocks to give brief speeches to the various ethnic groups clustered along the route of that polyglot melting pot—Ukrainians, Russians, Jews, Poles, Armenians, French, Czechs, Greeks, Hungarians, and Chinese. At one spot the First Army Band played "The Sidewalks of New York," and at another stop a Chinese fife-and-drum corps followed the parade with dancers and the clang of cymbals. The el was going out in style.

Soon sharp-tongued torches and grinding cranes began to eat away at the old el structure. Blue smoke drifted around the sloping staircases, and jackhammer crews beat a staccato at the base of the picturesque old gingerbread stations. The pillars were hard to pull up, because each was buried in a massive concrete base a yard below the surface and rooted in a two-inch plate anchor. They were stubborn—like wisdom teeth—requiring strenuous extraction. Chatham Square was exposed to the sunlight for the first time in three quarters of a century. Up the line the

El riders could experience the thrill of meandering down through the man-made canyons of lower Manhattan. South Ferry station at left. Third Avenue el looping through Water Street on right. 1941. PHOTO BY ARTHUR ROTHSTEIN, COURTESY PRINTS AND PHOTOS DIVISION, LIBRARY OF CONGRESS.

Third Avenue el. View north from the lower end of the uptown station at Fifty-ninth Street. Bloomingdale's Department Store is just across the tracks. 1951. COURTESY THE SMITHSONIAN INSTITUTION.

vanishing elevated opened to the sunlight a thousand stores, barber shops, restaurants, cheap hotels, hock shops, and barrooms. Shopkeepers stepped outside and blinked at the unaccustomed sunshine.

For a time Lipsett, the demolition contractor, was besieged by souvenir hunters and nostalgiacs, who wanted a trophy to remember the el by—a leather strap, a rusted spike, a blue glass station sign, potbellied stove, or turnstile. Some NBC executives wanted spikes to have chrome-plated for paperweights. One man wanted a batch of crossties, and a rich country matron from Pennsylvania tried to get the wreckers to take a whole station apart for reassembly at her estate. Lipsett refused.

The East Side looked strangely different. With its sprawling railroad trestles gone and the sun pouring in, the Bowery began to look like just another broad street and soon began to lose its time-blackened slums in favor of sunnier, more

*Coming down. A traveling crane lowers a heavy section
of steel to the street after its supporting pillars have
been cut away with torches.* COURTESY THE SMITHSO-
NIAN INSTITUTION.

Looking downtown from the platform at Third Avenue and Fifty-third Street, 1942. Chrysler Building in the distance. PHOTO BY MARJORY COLLINS, COURTESY PRINTS AND PHOTOS DIVISION, LIBRARY OF CONGRESS.

domestic life of the city—second-story apartment windows, washlines, backyards, sidewalks, and saloons. The el gave a traveler a special vantage point, not terribly high, but high enough—a unique vista in motion along earth-bound buildings. And the speed was slow enough for snooping, yet fast enough not to embarrass the voyeur or the viewed.

Though the el created a slum in places, it also created a permanent side show. Time was when children were taken a ride on it as a treat, something like Coney Island. Mannevillette Sullivan of Washington, D. C., who used to visit her uncle on Riverside Drive in the 1930s said, "A ride on the el was the high spot of my visit; you felt like you were in the clouds."

In time the elevated became old enough, remote enough, and bizarre enough to be nostalgic, quaint, and artistic. Not only was it a memorable example of nineteenth-century design; it was also a monument in the great tradition of ironwork construction. Like the Eiffel Tower or the Crystal Palace it was built of small, structural parts used visually and ornamentally with a surprising degree of aesthetic quality and craftsmanship—so far different from today's omnipresent slick, skin-covered architecture and one that is getting difficult to find.

Underneath the latticework of the superstructure it was like a roofed corridor, not really such a bad place. And the sun managed to filter through the bower overhead, creating a pattern of light and shadow that enchanted many artists. In the 1930s and after, the el was a favorite subject of such prominent painters as Edward Hopper, John Sloan, Louis Guglielmi, W. P. Snyder, and Jan Lechay. The old el was especially fascinating to Reginald Marsh, who made use of Third Avenue el atmosphere on many canvases. In 1954 Marsh was chosen to receive the Gold Medal of the National Institute of Arts and Letters, the highest award in the American cultural world. Interviewed when the award was announced, Marsh sounded an elegiac note. Everything, he said, reminded him of "the wonderful old world that's passing, the rococo boxes in the burlesque theaters, the elevated structure, the dark, drab streets underneath, tha tattooing shops, the cheap hotels and flophouses. All the things of the old days were so much better to draw. I hate to go down to Chatham Square where they're tearing down the span. The light and shadows on the characters

respectable abodes. The Bowery, where the el once roared overhead like a bursting dam, is quieter now, though the rumble of the trucks and buses seems oddly enough louder than before now that the shadowed cover of the el was removed. Today it has become a wide, busy avenue, though many of its former inhabitants still cling to their familiar haunts.

The change in Third Avenue since the demolition of the el has been even greater. After 1956 real estate values went bullish, and the area seemed to bloom. Row after row of dingy walkups have been replaced by high glass office buildings, luxurious apartments, and modern shops that present a solid, white front to the street. The character of the avenue has undergone a renaissance, and living and business are much brighter than before.

The el is gone. It is now only a memory, an item in history like the Blizzard of 1888 or the sinking of the *Titanic*. But there are those who mourn its passing, for it was always an adventure to ride on. Though it was an anathema to the motorist, the city planner, the architect, and the noise-crusader, it was nevertheless a delight to the passenger who could travel in the open air enjoying a cool breeze in the summer and looking idly down on the

Third Avenue train coming into Chatham Square Station, October 1951. COURTESY THE NEW YORK HISTORICAL SOCIETY.

walking around down there—you can't find any-thing better to draw. All these aspects of an older New York are out of date. Nobody wants them.''

Now that the invisible subway and the transient motorbus have replaced the el, the city has lost an important quality of street architecture, for in addition to being a means of transportation, it was meant to be looked at and to give pleasure in being seen. It was a welcome variation to the plain and naked city streets. The el is a reminder that emotional as well as practical considerations are necessary to make the ride to the office a pleasure rather than an ordeal.

It seems doubtful that elevated railways will ever again enjoy the vogue they once experienced. However, the same idea translated into contempo-rary terms has become commonplace. Elevated highways are familiar sights in cities all over the world, and such a scheme, if called a monorail instead of an el, seems daring and modern. Oddly enough, last summer the *Times* suggested that New York City solve its pollution and transit problems by bringing back Rufus Gilbert's clean and silent air-pressure elevated, complete with the colossal Gothic arches spanning the street. And who knows, now that the energy shortage seems to have ended the easy days of the private au-tomobile we may infact see such a revival.

Index